Impacts of Major Point and Non-Point Sources on Raw Water Treatability

The mission of the Awwa Research Foundation (AwwaRF) is to advance the science of water to improve the quality of life. Funded primarily through annual subscription payments from over 1,000 utilities, consulting firms, and manufacturers in North America and abroad, AwwaRF sponsors research on all aspects of drinking water, including supply and resources, treatment, monitoring and analysis, distribution, management, and health effects.

From its headquarters in Denver, Colorado, the AwwaRF staff directs and supports the efforts of over 500 volunteers, who are the heart of the research program. These volunteers, serving on various boards and committees, use their expertise to select and monitor research studies to benefit the entire drinking water community.

Research findings are disseminated through a number of technology transfer activities, including research reports, conferences, videotape summaries, and periodicals.

Impacts of Major Point and Non-Point Sources on Raw Water Treatability

Prepared by:
Grantley W. Pyke and **William C. Becker**
Hazen and Sawyer, P.C., New York, NY 10018
and
Richard Head
WRc/D&B, LLC, Trevose, PA 19053
and
Charles R. O'Melia
Johns Hopkins University, Baltimore, MD 21218

Jointly Sponsored by:
Awwa Research Foundation
6666 West Quincy Avenue, Denver, CO 80235-3098
and
Water Environment Research Foundation
601 Wythe Street
Alexandria, VA 22314

Published by:

Awwa Research Foundation
Advancing the Science of Water®

IWA Publishing

DISCLAIMER

This study was funded by the Awwa Research Foundation (AwwaRF). AwwaRF assumes no responsibility for the content of the research study reported in this publication or for the opinions or statements of fact expressed in the report. The mention of trade names for commercial products does not represent or imply the approval or endorsement of AwwaRF. This report is presented solely for informational purposes.

Published by IWA Publishing, Alliance House, 12 Caxton Street, London SW1H 0QS, UK

Telephone: +44 (0) 20 7654 5500; Fax: +44 (0) 20 7654 5555; Email: publications@iwap.co.uk;
Web: **www.iwapublishing.com**

AwwaRF report number 90959F. Originally published by AwwaRF for its subscribers in 2003.
IWA Publishing version published 2004.

ISBN 1 84339 866 4

CONTENTS

TABLES

FIGURES

FOREWORD

The Awwa Research Foundation is a nonprofit corporation that is dedicated to the implementation of a research effort to help utilities respond to regulatory requirements and traditional high-priority concerns of the industry. The research agenda is developed through a process of consultation with subscribers and drinking water working professionals. Under the umbrella of a Strategic Research Plan, the Research Advisory Council prioritizes the suggested projects based upon current and future needs, applicability, and past work; the recommendations are forwarded to the Board of Trustees for final selection. The foundation also sponsors research projects through the unsolicited proposal process; the Collaborative Research, Research Applications, and Tailored Collaboration programs; and various joint research efforts with organizations such as the U.S. Environmental Protection Agency, the U.S. Bureau of Reclamation, and the Association of California Water Agencies.

This publication is a result of one of these sponsored studies, and it is hoped that its findings will be applied in communities throughout the world. The following report serves not only as a means of communicating the results of the water industry's centralized research program, but also as a tool to enlist the further support of the nonmember utilities and individuals.

Projects are managed closely from their inception to the final report by the foundation's staff and large cadre of volunteers who willingly contribute their time and expertise. The foundation serves a planning and management function and awards contracts to other institutions such as water utilities, universities, and engineering firms. The funding for this research effort comes primarily from the Subscription Program, through which water utilities subscribe to the research program and make an annual payment proportionate to the volume of water they deliver and consultants and manufacturers subscribe based on their annual billings. The program offers a cost-effective and fair method for funding research in the public interest.

A broad spectrum of water supply issues is addressed by the foundation's research agenda: resources, treatment and operations, distribution and storage, water quality and analysis, toxicology, economics, and management. The ultimate purpose of the coordinated effort is to assist water suppliers to provide the highest possible quality of water economically and reliably. The true benefits are realized when the results are implemented at the utility level. The foundation's trustees are pleased to offer this publication as a contribution toward that end.

Edmund G. Archuleta. P.E.
Chair, Board of Trustees
Awwa Research Foundation

James F. Manwaring, P.E.
Executive Director
Awwa Research Foundation

ACKNOWLEDGMENTS

The authors thank the Awwa Research Foundation and the Water Environment Research Foundation for providing the opportunity to study an issue of great importance to the drinking water treatment community. The authors also extend their appreciation to AwwaRF project manager Linda Reekie for her guidance and support, and to the following members of the Project Advisory Committee for their insight and thoughtful review:

- Michael Jawson, U.S. Department of Agriculture, Agricultural Research Service
- Dolan McKnight, North Texas Municipal Water District
- Michael Pickel, Philadelphia Water Department
- Perri Standish-Lee, Black and Veatch
- John Thorne, Capitolink

Several utilities played an integral role in this project by contributing many hours of staff time collecting data to support the water treatment plant modeling task. The authors extend their profound thanks to the following supporting utilities and personnel:

- Fairfax County Water Authority, Fairfax, VA: Tom Bonacquisti & Melissa Billman
- Indiana American Water Company, Muncie, IN: Jeff Robinson and Christian Volk
- Millwood Water Treatment Plant: Gerard Moerschell (Town of New Castle, NY); David Rambo and Stephan Schmidt (Aquarion Services)
- Onondaga County Water Authority, Syracuse, NY: Anthony Geiss and Bob Rusyn

The authors also gratefully acknowledge the significant contributions of several other project participants. John O'Melia deserves credit for helping craft the overall project approach, as well as for his work with the agricultural best management practices databases. Deb Caraco and other staff at the Center for Watershed Protection played a critical role in modifying the Watershed Treatment Model for use in this project. Kimberly Davis of Hazen and Sawyer provided a helpful review of urban BMP findings. Richard Gullick of the American Water Works Service Company consistently facilitated the water quality data collection process. Jeremy Dudley of WRc provided faithful technical support during the water treatment plant modeling phase. Finally, Lyn Kirschner, Dan Towery, and others at the Conservation Technology Information Center reviewed the agricultural BMP performance data.

EXECUTIVE SUMMARY

Point and nonpoint source pollution of surface waters has a negative impact not only on the health of aquatic ecosystems but also on the drinking water treatment plants (WTPs) that rely on surface waters. Sediment, nutrients, pesticides, pathogens, and other pollutants in source waters can decrease treatability, increase treatment costs, and ultimately increase risks to public health. Utility managers typically respond to deteriorating raw water quality by increasing chemical dosages, modifying existing processes, or adding additional processes. As an alternative or supplement to changes at the treatment plant, managers may also consider promoting best management practices (BMPs) in the watershed to preserve or improve raw water quality. Given limited resources, however, managers need guidance on allocating resources between increased treatment at the plant and promotion of BMPs in the watershed.

RESEARCH OBJECTIVES

The purpose of this project is to guide utility managers and water quality personnel in determining where their resources should be focused by assessing the effects of watershed activities on raw water quality and water treatment costs. The primary objectives of this project are to:

1. Evaluate the effects of major point and nonpoint pollutant sources on raw water quality, treatability, and treatment costs
2. Evaluate the potential of agricultural and urban BMPs to mitigate these effects
3. Compare the efficacy and cost of BMPs in the watershed with increased treatment at the WTP

Five pollutants were selected for consideration in this project based on their frequency of occurrence and their impact on raw water treatability, treatment costs, or public health: total suspended solids (TSS), total organic carbon (TOC), total phosphorus (TP), *Cryptosporidium*, and atrazine. Though there are clearly other contaminants that may have a serious impact on utilities, these parameters represent some of the major challenges facing many water utilities, including treatability, nutrient over-enrichment, aesthetic quality, disinfection byproducts (DBPs), pesticides, and resistant pathogens.

APPROACH

In order to provide an assessment of the impact of watershed activities on WTP operations and costs that is applicable to a wide array of water treatment plants and watersheds, a loosely coupled watershed – water treatment plant modeling framework was developed. A screening-level watershed model (Watershed Treatment Model, Version 3.0, Caraco 2001) was used to assess the impact of watershed influences such as population growth, land use change, and BMP implementation on pollutant export to receiving waters. A dynamic water treatment plant simulation model (OTTER, Version 2.1.2, WRc/D&B 2001) was then used to assess the impact of raw water quality changes on the operations, costs, and finished water quality of different types of treatment plants. Cost-benefit analysis was conducted to estimate the percentage of WTPs that would find BMP implementation to be a cost-effective means of reducing routine operating costs.

Contaminants of concern included in the watershed modeling include TSS, TP, and TOC. Parameters included in the WTP modeling include TSS, TOC, and atrazine. Cost-benefit analysis was conducted for TOC and TSS. Due to model constraints, *Cryptosporidium* and atrazine were not included in the watershed modeling, and TP and *Cryptosporidium* were not included in the WTP modeling. Qualitative assessment was provided for these parameters.

RESULTS

Major Point and Nonpoint Pollutant Sources

Export coefficient-based watershed modeling using the Watershed Treatment Model (WTM) was conducted to investigate the relative contribution of point and nonpoint pollutant sources to total watershed export of TOC, TP, and TSS. Modeling scenarios were based on hypothetical watersheds with land use and population density characteristic of typical forested, agricultural, and mixed land use drinking water supply watersheds. Major findings for these watersheds include:

- The majority of TOC, TP, and TSS exported to receiving waters comes from nonpoint sources. The relative contribution of different nonpoint sources depends heavily on the fraction of total watershed area devoted to a given land use. In forested watersheds, forest cover is the dominant source of TOC and a major source of TP and TSS. In agricultural watersheds, cropland is the dominant source of TOC, TP, and TSS. In the mixed land use watersheds modeled, forest and cropland are major sources of TOC and TSS, and cropland and point sources (i.e. wastewater treatment plants) are the dominant sources of TP.
- The primary water quality impact of pasture and grazing land is associated not with surface runoff but with direct livestock access to streams and stream banks, which increases sediment, nutrient, and pathogen loads to receiving waters.
- Runoff from concentrated animal feeding operations (CAFOs) depends on site-specific waste-handling systems. Land application of CAFO waste at non-agronomic rates is common and leads to phosphorus buildup in soils and increased phosphorus runoff. Model predictions indicate that CAFOs of all types are expected to make significant contributions to TOC and TP export in areas where manure storage and treatment is inadequate and manure application rates exceed crop utilization rates.
- For the watersheds modeled in this project, pollution from urban land use alone, even at relatively high population densities, contributes less than 10% of the total watershed export of TOC, TP, and TSS. For the pollutants considered in this project, the major impacts of population growth and development are increased TP loads from wastewater treatment plants and increased TOC, TP, and TSS loads from channel erosion caused by increased impervious cover.
- Model predictions using best available load estimation parameters indicate that failing septic systems, CSOs, and SSOs are not typically major contributors to total annual export of TOC, TP, and TSS.

Potential of BMPs to Reduce Pollutant Loads

The Watershed Treatment Model was used to investigate the potential of BMPs to reduce total watershed export of TOC, TP, and TSS. Modeling scenarios were based on typical pollutant removal efficiencies for common urban and agricultural BMPs. Findings regarding the potential of BMPs to reduce pollutant loads include:

- Agricultural BMPs such as buffer strips and conservation tillage can reduce the export of TP, TSS, and pesticides from cropland, though the magnitude of the reduction depends heavily on the fraction of total cropland treated by BMPs. The impact of agricultural BMPs on TOC export is not well characterized and merits further research.
- Improved animal waste management at CAFOs and pasture/range BMPs such as livestock exclusion can reduce TOC, TP, and TSS loads in watersheds with intensive animal production.
- Because urban land use is expected to contribute a small fraction of total TOC, TP, and TSS loads in most drinking water supply watersheds, urban BMPs have limited potential to reduce total export of these parameters. Urban BMPs are expected to provide significant reductions of TOC, TP, and TSS only for WTPs treating source waters heavily impacted by highly urbanized subwatersheds.
- The long-term performance of most urban and agricultural BMPs is heavily dependent on effective maintenance.

Impact of Pollutant Loads on Water Treatment Plants

Dynamic water treatment plant modeling was conducted to investigate the effects of changes in raw water TOC and TSS concentrations on WTP operations, costs, and finished water quality. Modeling scenarios were based on four different treatment plant types treating four different source waters. Operating costs quantified include chemical consumption, residuals disposal, and filter backwashing. Findings regarding the impact of pollutant loads on treatment costs include:

- Increases in raw water TOC and TSS concentrations increase WTP operating costs due to increased residuals disposal, filter backwashing, and chemical consumption. Moderate deteriorations in raw water quality (e.g. 25% increase in solids and 25% increase in organics) can increase treatment costs by roughly 10%. Moderate improvements (e.g. 25% decrease in solids and 15% decrease in organics) can reduce treatment costs by roughly 5%. Operating costs do not exhibit high sensitivity to changes in TOC and TSS concentrations (that is, the percent change in operating costs is less than the percent change in pollutant loads). The observed changes in operating costs are nonetheless significant.
- Increases in point and nonpoint source phosphorus loads accelerate the natural process of eutrophication in surface waters, which decreases treatability, increases treatment costs, and may increase concentrations of DBP precursors. The magnitude of these changes is highly site-specific and is difficult to quantify on an average basis.

- Export of agricultural pesticides to receiving waters depends heavily on application methods and timing relative to rainfall events. Raw water atrazine concentrations in excess of the current 3 ppb finished water maximum contaminant level (MCL) are common after spring rainfall events in high atrazine-use watersheds. Treatment plants in these watersheds may incur substantial additional treatment costs in order to comply with water quality standards.
- Under the expected terms of the Long Term 2 Enhanced Surface Water Treatment Rule (LT2ESWTR), WTPs drawing from raw water sources with *Cryptosporidium* concentrations greater than 0.075 oocysts/L will be required to provide additional treatment. Depending on raw water oocyst levels, watershed protection measures may satisfy some or all of these additional treatment requirements. Treatment plants drawing from source waters containing more than 1 oocyst/L are expected to incur substantial costs for the addition of new processes such as ozonation, UV irradiation, or membrane filtration.

BMP Cost-Effectiveness

Findings regarding BMP cost-effectiveness include:

- Targeting BMPs to priority pollutant source areas is critical for maximizing the impact of source protection funds. Many BMP programs do not target BMP funding to the most significant pollutant sources in the watershed.
- BMP cost-effectiveness depends in large part on the scale of the watershed protection effort and the fraction of total BMP implementation cost that is shared by the utility. WTPs would seldom benefit from funding 100% of BMP implementation as a means of reducing routine operating costs. However, most watershed protection efforts involve multiple partners and funding sources. Utilities may be able to improve or protect source water quality and reduce treatment costs at minimal direct expense by forming source water protection partnerships with federal, state, and local interests. Utility participation in watershed protection efforts, even at zero or low BMP cost-share levels, can help prioritize allocation of state and federal funding sources for water supply protection.
- Promotion or cost-sharing of BMPs such as conservation tillage and riparian buffer strips is expected to be a cost-effective strategy for reducing operating costs for some utilities in watersheds heavily influenced by row-crop agriculture. An analysis of national data on the size of watershed size relative to treatment plant size was used to provide a screening-level estimate of the fraction of water treatment plants that would benefit from funding BMP implementation. Assuming that WTPs fund 10% of the total BMP cost, and that other funding sources are available to provide sufficient cost-share funding to achieve BMP implementation, the fraction of WTPs benefiting from BMP implementation ranged from 5 - 15% for riparian buffers to 70 – 80% for conservation tillage.
- Structural urban BMPs are expected to be a cost-effective strategy for reducing operating costs only for WTPs treating source waters with pollutant loads derived

primarily from urban NPS pollution. The long-term effects of population growth and increased impervious cover in urbanizing watersheds are best countered at an early stage through promotion of planning and development practices that reduce impervious cover and prevent channel erosion.

- Source water protection measures that decrease concentrations of DBP precursors, pathogens, and pesticides in source waters have public health benefits that are not adequately expressed in strict economic terms.
- The modeling scenarios in this report describe the effect of changes in TOC and TSS loads on routine operating costs. They do not quantify the effects of other pollutants on WTP costs, nor do they account for the costs of major process changes or source of supply changes caused by severe water quality degradation. Due to the high cost of WTP capital improvements relative to watershed BMPs, utilities should consider promoting long-term source water protection measures that have the potential to prevent major process changes.
- Overall, this report is believed to present a conservative assessment of BMP cost-effectiveness.

Watershed-Water Treatment Plant Modeling Approach

A coupled watershed-water treatment plant modeling approach provides a useful framework for utility managers seeking to assess the costs and benefits of source water protection measures. Observations regarding the modeling tools applied in this project include:

- The watershed model used in this project (the Watershed Treatment Model) provides a useful screening-level assessment of the relative importance of different point and nonpoint sources on annual export of TOC, TP, and TSS. It also indicates the magnitude of pollutant load reductions that can be expected with BMP implementation. The model does not simulate fate and transport of contaminants within the watershed, nor does it predict pollutant export from pervious land cover during runoff events.
- In order to provide an assessment of watershed changes on water treatment plants that was not site-specific, pollutant fate and transport in receiving waters was not modeled in this project. Dynamic, site-specific receiving water modeling is required to provide a more accurate account of the effects of pollutant attenuation and transformation in the receiving water on raw water quality at the treatment plant intake. This is particularly true for sediment, phosphorus, and organic carbon. More sophisticated receiving water modeling is required to link changes in watershed export of phosphorus with water quality parameters of interest at WTPs.
- The water treatment plant model used in this project (OTTER) is a powerful tool that can effectively simulate most water treatment plant processes and parameters of concern. It has proved useful in this project for assessing the impact of changes in raw water TOC and TSS concentrations on WTP operations, costs, and finished water quality. However, the model is data and labor-intensive and may exceed the resource levels of many small or medium-sized utilities.

RECOMMENDATIONS

Determining whether it makes economic sense for a utility to invest in watershed protection as a means of reducing current and future treatment costs is a complex and site-specific decision-making process. Recommendations for utilities considering source protection include:

- Use the Source Water Assessment Plan to identify significant potential sources of your contaminants of concern within your watershed. When developing a source water protection strategy, target BMP promotion to these areas.
- For utilities in small watersheds, use a simple watershed model such as the WTM to develop a planning-level estimate of the major pollutant sources in your catchment.
- For all utilities, partner with existing watershed protection or conservation groups in order to emphasize the importance of protecting raw water supplies. Utility participation in watershed protection efforts can help prioritize allocation of BMP funds.
- Start addressing urban NPS pollution now. The effects of population growth are difficult and expensive to mitigate after the fact. Low-cost efforts to reduce increases in new impervious cover through promotion of zoning ordinances and low-impact development practices will help reduce the potential for long-term water quality deterioration.
- Address agricultural NPS pollution from crop and pasture land by partnering with farm groups, county-level Soil and Water Conservation Districts, or extension agents. These stakeholders have considerable knowledge of significant pollutant sources within the watershed, and are well acquainted with BMPs and funding sources. Consider cost-sharing a small percentage of BMP implementation as a means of leveraging additional funding sources and reducing treatment costs.
- Address pollutant loads from CAFOs by promoting improvements in agricultural waste management systems, as well as land application of animal waste at phosphorus-based rates.
- When evaluating investment in source protection measures, consider the long-term impact of these measures on preventing water quality degradation and reducing future treatment costs and facilities requirements. Consider as well the non-economic benefits of source protection, including the potential public health benefits of reduced exposure to pesticides, pathogens, and emerging contaminants of concern.

FUTURE RESEARCH

The following research needs were identified during the course of this study:

- Better characterization of the DBP formation potential of different TOC sources is required to improve our understanding of the effects of watershed activities on DBPs in drinking water. TOC is measured less frequently than nutrients and sediment in BMP studies and ambient water quality monitoring, and the effect of most BMPs on the export of TOC and DBP precursors is poorly characterized. This is particularly important for BMPs such as conservation tillage and riparian buffers that increase plant biomass and soil organic matter carbon levels. These BMPs are widely recommended and implemented; yet, their effect on DBP precursor export is poorly understood. Similarly, the export of pathogens and DBP precursors from concentrated animal feeding operations using different types of waste handling and disposal systems merits further research.

- Improved modeling and monitoring of the effects of population growth and land development on pollutant loads from channel erosion is necessary to guide source protection efforts in developing watersheds
- Improved modeling and monitoring of the effects of BMPs on *Cryptosporidium* export from agricultural watersheds will enhance the effectiveness of source protection programs implemented under the LT2ESWTR.
- A better national-scale assessment of the size, land use, and population density of water supply watersheds will help guide source water protection policy, funding, and program implementation.

CHAPTER 1
INTRODUCTION

Point and nonpoint source pollution of surface waters has a negative impact not only on the health of aquatic ecosystems but also on the drinking water treatment plants (WTPs) that rely on surface waters. Sediment, nutrients, pesticides, pathogens, and other pollutants in source waters can decrease treatability, increase treatment costs, and ultimately increase risks to public health. Utility managers typically respond to deteriorating raw water quality by increasing chemical dosages, modifying existing processes, or adding additional processes. As an alternative to changes at the treatment plant, managers may also consider investing in source water protection measures in order to preserve or improve raw water quality. Given limited resources, however, managers need guidance on how to allocate funds between watershed protection and increased treatment at the plant.

PURPOSE AND SCOPE

The purpose of this project is to guide utility managers and water quality personnel in determining where their resources should be focused by assessing the effects of watershed activities on raw water quality and water treatment costs. The primary objectives of this project are to:

1. Evaluate the effects of major point and nonpoint pollutant sources on raw water quality, treatability, and treatment costs
2. Evaluate the potential of agricultural and urban BMPs to mitigate these effects
3. Compare the efficacy and cost of BMPs in the watershed with increased treatment at the WTP

This report is organized as follows. First, the modeling methodology and the importance of the selected contaminants of concern are described below. A discussion and literature review of the impact of agricultural NPS pollution and the mechanisms and efficiencies of selected agricultural BMPs are presented in Chapter 2. A detailed description of the methods used for the watershed modeling portion of this project is presented in Chapter 3, along with a literature review regarding the subject, while watershed modeling results are presented in Chapter 4. Methodology and results of water treatment plant modeling are presented in Chapters 5 and 6. The cost-benefit analysis methodology and results are presented in Chapter 7. Sources and impacts of phosphorus and *Cryptosporidium* are discussed in more detail in Chapters 8 and 9. Lastly, Chapter 10 summarizes research findings, while recommendations to water utilities are presented in Chapter 11.

CONTAMINANTS OF CONCERN

Though water utilities are faced with a seemingly ever-expanding array of contaminants of potential concern, in practice many utilities are challenged by a relatively small group of pollutants. Five pollutants were selected for consideration in this project based on their frequency of occurrence and their impact on raw water treatability, treatment costs, or public health:

- Total suspended solids (TSS)
- Total phosphorus (TP)
- Total organic carbon (TOC)
- Atrazine
- *Cryptosporidium*

Though there are clearly other contaminants that may have a serious impact on utilities, these parameters were chosen in consultation with project advisors to represent some of the major challenges facing many water utilities, including treatability, nutrient over-enrichment, aesthetic quality, disinfection byproducts, pesticides, and resistant pathogens.

Total suspended solids (TSS) is included because of its impact on coagulation, clarification, and filtration processes. Increased sediment loads increase operating costs through increased oxidant and coagulant demand and residuals disposal requirements. Peak sediment loads during runoff events can decrease treatability, increase clarified turbidity, and decrease filter run time. Shorter filter runs mean more frequent backwashing, reduced treatment capacity, and increased power consumption. Peak solids loads also drive the sizing of clarification processes, so increases in solids export during runoff events may increase the size and cost of new unit processes. At direct filtration plants, increased solids loading rates may ultimately require the addition of a clarification process. For treatment plants withdrawing from lakes or reservoirs, excess sediment loads decrease raw water storage capacity and may increase dredging frequency.

Total organic carbon (TOC) has a critical effect on coagulation performance as well as subsequent treatment processes. Increased TOC loads can be expected to increase operating costs by increasing oxidant and coagulant demand. TOC is also an indicator of a source water's potential to form disinfection byproducts (DBPs). Increased TOC loads may increase concentrations of trihalomethanes and haloacetic acids in finished water. Elevated DBP levels have a negative public health impact, and compliance with maximum contaminant levels (MCLs) may require treatment process changes and increase treatment costs.

Though phosphorus in drinking water is not a direct health concern, it is included as a contaminant in this research because elevated phosphorus concentrations can accelerate the natural process of eutrophication and lead to a variety of operational and aesthetic water treatment problems. As the limiting nutrient in most fresh waters, phosphorus stimulates the growth of aquatic biota such as algae. Excessive algal growth may trigger taste and odor episodes, severe pH fluctuations, release of iron and manganese from sediments, and filter clogging in treatment plants. In addition, increased primary productivity and autochthonous organic carbon production can lead to an increase in raw water DBP formation potential and subsequent negative health impacts.

Cryptosporidium oocysts are a critical concern to water utilities due to their presence in surface waters, their resistance to conventional disinfection, and their public health significance. *Cryptosporidium* oocysts can cause cryptosporidiosis, a diarrheal disease that is potentially life threatening to individuals with underdeveloped or compromised immune systems. Depending on the concentration, *Cryptosporidium* in source waters can trigger capital expenditures for installation of new processes such as ozonation, UV disinfection, or membrane filtration.

Since many pesticides are minimally removed by conventional treatment processes, elevated concentrations of pesticides in source waters may increase risks to public health, and compliance with MCLs can increase treatment costs. Though MCLs have not been established for many pesticides, concentrations of commonly used agricultural herbicides such as atrazine have increased treatment costs for some utilities by increasing operating costs (e.g. increased use of powdered activated carbon) or requiring the installation of new processes such as granular activated carbon filter media. Atrazine was selected as a representative pesticide in this project because it is the most widely used herbicide in the U.S. Atrazine has been applied to corn, sorghum, sugar cane, and other crops for over 40 years, and is used throughout much of the Midwestern U.S. Atrazine was detected in 85% of stream samples collected from 62 agricultural watersheds from 1992 – 1998 (Larson 2001). Atrazine is not removed through conventional treatment processes, and recent studies indicate that it disrupts endocrine system function (see e.g. Cooper et al. 2000).

METHODOLOGY

In order to provide an assessment of the impact of watershed activities on WTP operations and costs that is applicable to a wide array of water treatment plants and watersheds, a loosely coupled watershed – water treatment plant modeling framework was developed. A watershed model (Watershed Treatment Model, Version 3.0) was used to assess the impact of watershed influences such as population growth, land use change, and BMP implementation on pollutant export to receiving waters (Caraco 2001). A water treatment plant simulation model (OTTER, Version 2.1.2) was then used to assess the impact of changing raw water quality on water quality, operations, and treatment costs (WRc/D&B 2001). Contaminants of concern included in the watershed modeling include TSS, TP, and TOC. Parameters included in the WTP modeling include TSS, TOC, and atrazine. Due to model constraints, *Cryptosporidium* and atrazine were not included in the watershed modeling, and TP and *Cryptosporidium* were not included in the WTP modeling.

For the parameters covered in both the watershed and WTP models (TOC and TSS), a cost-benefit analysis was conducted to (1), assess the impact of increased or decreased loads of these constituents on WTP operating costs; and (2), assess the cost-effectiveness of BMP implementation as a means of reducing treatment costs. Tools used to investigate each of the parameters of concern are summarized in Table 1.1 below. Additional detail on the Watershed Treatment Model, OTTER, and the interaction between these models is provided below.

Table 1.1
Contaminants of concern and assessment tools

	Literature Review	Watershed Modeling	Water Treatment Plant Modeling	Cost-Benefit Analysis
Total suspended solids	✔	✔	✔	✔
Total phosphorus	✔	✔	-	-
Total organic carbon	✔	✔	✔	✔
Atrazine	✔	-	✔	-
Cryptosporidium	✔	-	-	-

The Watershed Treatment Model (WTM) is a screening-level, export-coefficient based watershed model that was used to quantify the relative changes in annual pollutant loads exported from sets of hypothetical watersheds under varying land use and BMP implementation scenarios. Pollutant sources simulated include major land uses (urban, forest, cropland, and pasture), wastewater treatment plants (WWTPs), combined sewer overflows (CSOs), sanitary sewer overflows (SSOs), failing septic systems, concentrated animal feeding operations (CAFOs), and channel erosion. Agricultural BMPs considered in this project include conservation tillage, buffer strips, settling basins, wetlands, nutrient management, grazing management, and animal waste management. Structural urban BMPs covered include buffer strips, dry and wet ponds, wetlands, swales, filters, and infiltration trenches. Watershed modeling results were expressed as the percent change in annual pollutant export between a base case scenario and an alternative scenario. Watershed modeling results are presented for TSS, TOC, and TP.

Watershed model output was then used as input for the OTTER water treatment plant model. OTTER was used to simulate the impact of changing raw water quality on the operations, treatment costs, and finished water quality of four types of treatment plants treating four different source waters. Treatment plants modeled include a conventional plant, a conventional plant with GAC filter media, a dissolved air flotation plant with ozone disinfection, and a direct filtration plant. Models were constructed, calibrated, and validated using plant physical characteristics, operations information, and water quality data. Water treatment plant modeling results are presented for TSS, TOC, and atrazine.

Coupled watershed-water treatment modeling was performed for TSS and TOC. In order to allow for a generalized assessment of the impacts of pollutants and BMPs on WTPs, transport and fate of these constituents in receiving waters was not explicitly considered. Instead, the watershed and water treatment plant models were linked by scaling daily values of turbidity and TOC in the OTTER raw water quality profiles up or down according to the relative change in annual TSS and TOC export in watershed model scenarios. For example, if the WTM indicates that annual TSS export increases by 25% from a base case scenario to an alternative scenario, then daily turbidity values in the OTTER base case raw water quality profile are increased by 25%. OTTER was then used to assess the impact of these changes in raw water quality on finished water quality and treatment plant operating costs. Cost parameters considered include chemical usage, filter backwash frequency, and residuals production.

Because the input to the water treatment plant model requires only the percent change in pollutant loads from the base case to an alternative scenario, the critical element of WTM output is not the magnitude of the pollutant load from any given watershed scenario, but rather the relative change in pollutant load between two scenarios. This reliance on relative differences helps to increase the generality of project findings by reducing the sensitivity of WTM output to the absolute value of model default pollutant export coefficients and pollutant runoff concentrations.

The generality of project findings is also increased by using hypothetical watersheds that represent typical land use characteristics for water supply watersheds. These characteristics are described further in Chapter 3. The impact of point sources such as wastewater treatment plants is investigated by adjusting key parameters such as population density to reflect nationally relevant values (e.g. median or other specified population density value). Similarly, the impact of concentrated animal feeding operations (CAFOs) is calculated by adjusting the livestock population of the watershed to reflect specified animal unit density values.

CHAPTER 2
AGRICULTURAL NONPOINT SOURCE POLLUTION AND BMPS

INTRODUCTION

This chapter presents an overview of agricultural nonpoint source (NPS) pollution and the best management practices (BMPs) that are commonly used to control it. The first section briefly describes the extent and impact of agricultural NPS pollution. The second section reviews agricultural sources of five contaminants of concern (sediment, phosphorus, organic carbon, pesticides, and pathogens), the mechanisms by which they are exported from the landscape, and the mechanisms by which they may be removed from agricultural runoff. The third section describes common agricultural BMPs and presents typical BMP removal efficiencies. Major pollutant prevention BMPs discussed include conservation tillage, nutrient management, integrated pest management, grazing management, and animal waste management systems. Major runoff treatment BMPs discussed include buffers, settling basins, and wetlands.

EXTENT AND IMPACT OF AGRICULTURAL NPS POLLUTION

Awareness of the extent of both urban and agricultural nonpoint source pollution has increased in recent years as pollutant loads from point sources have decreased and ambient water quality monitoring has increased. The U.S. Geological Survey's National Water-Quality Assessment (NAWQA) Program, which assesses surface and groundwater quality in major basins and aquifers across the U.S., shows that excess nutrient and pesticide loads from nonpoint sources are common. Based on NAWQA Phase I (1992 - 1996) results for 20 major basins across the U.S., total nitrogen concentrations in 61% of sampled streams exceed a national average natural background concentration of 1.0 mg/L. Likewise, 57% of streams (and 80% of streams in agricultural areas) exceed the EPA target level of 0.1 mg/L phosphorus for preventing nuisance plant growth. About 90% of the nitrogen and 75% of the phosphorus originate from nonpoint sources (USGS 1999).

Agricultural activities comprise a significant fraction of nonpoint source nutrient loads. Nowhere is this more evident than the Gulf of Mexico, where a seasonal 6,000 square mile hypoxic (oxygen-depleted) zone is attributed to high nutrient loads from the Mississippi-Atchafalaya River Basin. Roughly ninety percent of the nitrogen and phosphorus discharging to the Gulf comes from nonpoint sources. Primary sources of nitrate include the fertilizer-soil nitrate pool (58%), manure (16%), municipal and industrial point sources (9%), and other inputs such as atmospheric deposition, ground water, and soil erosion (16%). Intensively cultivated regions in southern Minnesota, Iowa, Illinois, Indiana, and Ohio contribute a disproportionately large amount of nitrogen to the Gulf: streams draining Iowa and Illinois (9% of the basin area) contribute up to 35% of the total nitrogen flux to the Gulf. The main phosphorus sources include fertilizer (31%), manure (17%), municipal and industrial point sources (10%), and other inputs such as soil erosion (42%) (Goolsby et al. 1999).

Agricultural activities have also been identified as a major source of pesticides in surface waters. Low levels of herbicides and insecticides were widely detected in the first phase of the NAWQA program, occurring in over 90% of samples of monitored streams (USGS 1999). Herbicide concentrations were highest in areas with high agricultural land use; insecticide concentrations were highest in urban streams. Herbicide detection frequency is closely tied to usage, with commonly used herbicides such as atrazine, metolachlor, alachlor, and cyanazine detected in 30 to 85% of sampled streams (Larson 2001).

The impact of agricultural nonpoint source pollution on surface water quality has received increasing attention. Nationwide, agricultural operations have been identified as a leading cause of water body impairment. Under Section 305(b) of the Clean Water Act, states and other jurisdictions assess and report to EPA the extent to which their waters support water quality standards and designated uses. In the 2000 305(b) summary report, which assessed 19% of the nation's stream miles and 43% of its lake and reservoir acres, agriculture was the most frequently identified known cause of impairment for both lakes and rivers. Agricultural sector operations were identified as a cause of water quality impairment in 48% of impaired rivers and 41% of impaired lakes. By comparison, far fewer impairments were attributed to municipal point sources (10% for lakes and 12% for rivers) and urban runoff/storm sewers (13% for lakes and 18% for rivers). Overall, 17% of stream miles and 22% of lake acres were classified as not fully supporting drinking water usage (USEPA 2002b).

Negative impacts of agricultural nonpoint source pollution on drinking water sources include increased treatment costs, decreased treatability, and potential negative health effects. Cost impacts vary widely depending on the contaminant of concern, its concentration and occurrence frequency, and the type and capacity of unit processes at the treatment plant. Several examples illustrating the impact of agricultural NPS pollution on treatment costs include:

- In 1990, City Water Light and Power of Springfield, Illinois completed an $8 million dredging project that restored roughly 500 million gallons of storage in its raw water source, Lake Springfield. Sediment loads from the 88% row-cropped watershed had reduced reservoir capacity at a rate of 50 million gallons per year. The 22 mgd utility also spends $50,000 - $200,000 annually on powdered activated carbon to reduce atrazine concentrations to acceptable levels (LSWRPC 1998).
- The Tulsa (Okla.) Metropolitan Utility Authority spends roughly $100,000 annually in an effort to alleviate severe taste and odor problems in its Eucha-Spavinaw supply (USEPA 2001f). Eutrophication of Lake Eucha is accelerated by large phosphorus inputs, roughly 75% of which are derived from nonpoint sources, primarily phosphorus runoff from excessive land-application of poultry manure (Storm et al. 2001).
- In response to seasonal nitrate levels in the Des Moines and Raccoon Rivers well above the 10 mg/L MCL, Des Moines (Iowa) Water Works installed in 1991 a $3.7 million ion exchange facility at its Fleur Drive treatment plant (average production 43 mgd). The facility is operated for one to two months per year at a cost of approximately $3000 per day. Row crops make up roughly 90% of the land use in the Raccoon River watershed and contribute an estimated 80% of the nitrate loads to the Raccoon River (DMWW 2002).
- In 1997, a $3.2 million powdered activated carbon facility was placed in service to reduce atrazine concentrations to acceptable levels at the Columbus (Ohio) Division of Water's Cremean WTP. The Cremean plant treats roughly 67 mgd of water from the Hoover Reservoir, which is fed by the 60% row-cropped Upper Big Walnut Creek watershed. Raw water atrazine levels peaked in the late 1980s

and early 1990s, when annual average raw water concentrations in excess of 3 ppb were recorded (Binder 2002).

While agricultural NPS pollution in intensively farmed watersheds has increased treatment costs for some water suppliers, it has potential negative health effects as well. USGS NAWQA data indicates that the current regulatory framework for pesticides does not address the typical pattern of exposure to pesticides in drinking water. Herbicide concentrations subsequent to spring runoff events can exceed drinking water MCLs for weeks in rivers and streams and months in lakes and reservoirs (see e.g. Capel et al. 2001 and Larson et al. 1999). However, compliance with MCLs is based on an annual average of quarterly samples and does not prevent such intermediate-term exposure to elevated pesticide concentrations. Furthermore, pesticide MCLs are established based on acute exposure to high doses of parent compounds, and do not account for chronic exposure to lower concentration mixtures of parent compounds and degradation products.

The EPA's revision of the atrazine risk assessment provides a rough indication of the extent of human exposure to atrazine via drinking water (USEPA 2001b). Based on quarterly SDWA monitoring data for 3,670 community surface water systems from 1993 - 1998, over 5 million people (7.9% of the assessed population) were supplied by a system that had detected atrazine at greater than 3 ppb, and over 1 million (1.7% of the assessed population) were supplied by a system with an annual average atrazine concentration greater than 3 ppb.[1] The database covers 21 major atrazine use states that represent 92% of the atrazine use in the U.S. It is important to note that this data does not represent peak atrazine exposure concentrations, nor does it indicate exposure to atrazine breakdown products.

CONCENTRATED ANIMAL FEEDING OPERATIONS

Animal feeding operations (AFOs) and concentrated animal feeding operations (CAFOs) are a potentially significant source of numerous pollutants, including nutrients, solids, pathogens, organic matter, metals, salts, antibiotics, and hormones. In general, AFOs and CAFOs keep animals in confinement and concentrate animals, feed, manure, dead animals, and production operations on a small land area. There are currently roughly 238,000 AFOs and 12,000 CAFOs in the U.S. (USEPA 2002a).

An animal operation is an AFO if it confines animals 45 or more days per year and if the confinement area does not support vegetation during the growing season. An animal operation is a CAFO if it meets the AFO definition and one of three CAFO definitions (USEPA 2002a):

- An AFO is a large CAFO if it confines a large number of animals (e.g. 1,000 beef cattle, 700 dairy cows, etc.).
- An AFO is a medium CAFO if it confines an intermediate number of animals (e.g. 300 – 1,000 beef cows, 200 – 700 dairy cows, etc.) and it either discharges animal waste to surface waters through a manmade ditch or pipe, or allows animal contact with surface water passing through the confinement area.

---

[1] Based on data from the Novartis PLEX database, as presented in USEPA 2001b.

- An AFO is a small CAFO if it confines fewer than a lower threshold number of animals (e.g. 300 beef cattle, 200 dairy cows, etc.), and it has been designated as a CAFO by the permitting authority.

Upper and lower size thresholds are established for cows, pigs, chickens, and other farm animals. Size thresholds depend on animal type, animal size, and waste handling systems. In general, any AFO regardless of size may be designated as a CAFO by the permitting authority if it is a significant contributor of pollutants to surface waters.

CAFOs are defined as point sources under the Clean Water Act, and have been regulated under the National Pollutant Discharge Elimination System (NPDES) program since 1976. Despite inclusion of CAFOs under the NPDES regulatory framework, evidence of the negative water quality impacts of CAFOs has increased since the 1970s. This evidence includes ecological impacts such as fish kills, eutrophication and algal growth, disturbance of aquatic ecosystems, as well as impacts on drinking water supplies and recreation areas. The extent of CAFO pollution is due in part to ineffective program implementation: of the 12,000 facilities currently defined as CAFOs, only 4,500 have applied for a NPDES permit (USEPA 2002a).

A more fundamental cause of increased CAFO pollution, however, is the tremendous consolidation that has occurred in the animal production industry over the past 25 years. While poultry and livestock production has increased significantly since the 1970s, the number of poultry and livestock farms in the U.S. declined by 40% from 1974 to 1997. Small and medium-sized growers have been replaced by large operations, a higher percentage of livestock is kept in confinement, and animal production has been linked more closely to processing facilities (USEPA 2001a). This consolidation of animal production has resulted in geographic concentrations of animal waste that exceed the nutrient assimilation capacity of the crop and pasture land that is available for manure disposal. The net effect is frequent land application of manure at non-agronomic rates and subsequent increases in nutrient export to receiving waters.

A study by the Natural Resources Conservation Service (NRCS) quantified the frequency of this problem (Kellogg et al. 2000). Authors used Agriculture Census data from 1982, 1987, 1992, and 1997 to compare the mass of N and P generated by livestock (beef cattle, dairy cattle, swine, and poultry) in a given county with the mass of N and P that could be assimilated by the cropland and pastureland in that county. The number of counties with manure surpluses more than doubled from 1982 to 1997.

Consolidation in the poultry and hog industry has been particularly pronounced, with manure production in excess of local nutrient requirements observed in the Delaware/Maryland/Virginia Peninsula and portions of North Carolina, Georgia, Alabama, Arkansas, Iowa, and Nebraska (USEPA 2001c). An examination of manure disposal records from 10 CAFOs housing 60,000 hogs in north-central Iowa indicated that land application of manure based on crop P demand would require 9,350 ha of cropland, compared to the 990 ha used (Jackson et al. 2000). Glasgow and Burkholder (2000) document over-concentration of phosphorus in North Carolina's Neuse River Basin due to growth of swine CAFOs.

While conditions such as nutrient over-enrichment are likely the most common chronic water quality impact of CAFOs, high-profile acute impacts such as lagoon spills or ruptures and piping failures also occur, often resulting in large fish kills and dramatic water quality degradation. Burkholder et al. (1997) document the effects of a large North Carolina swine waste spill on the receiving river and estuary. Sediment, nutrient, and fecal coliform concentrations in the plume were one to two orders of magnitude higher than baseline conditions, and fecal coliform concentrations in surface sediments remained above normal for at least 2 months in some locations. The spill was followed by increased phytoplankton production and decreased dissolved oxygen levels, as well as emergence of noxious algae and toxic dinoflagellates (*Pfiesteria piscicida*) that may have contributed to a large fish kill at the same time. Mallin et al. (1997) document similar effects from a swine waste spill and a chicken waste spill in North Carolina, and cite more than 30 waste lagoon failures in North Carolina in from 1995 to 1996.

In response to concerns about the negative health and environmental impact of animal feeding operations, EPA finalized the CAFO Rule in late 2002 (USEPA 2002a). The new CAFO Rule maintains the same basic structure and size thresholds that have been in place since 1976, but makes several significant modifications. First, all CAFOs are required to apply for a NPDES permit. This closes a loophole that exempted CAFOs which only discharged during 25-year, 24-hour storm events, and is expected to more than double the number of NPDES-permitted CAFOs. Second, large poultry operations are regulated. Third, all CAFOs covered by a NPDES permit are required to develop and implement a nutrient management plan. The nutrient management plan requirement will promote better manure management and attempt to prevent land application of manure at non-agronomic rates.

POLLUTANTS: SOURCES, TRANSPORT, AND REMOVAL MECHANISMS

Sediment

Sources of sediment in agricultural runoff include native soils, as well as amendments applied to cropland, pastureland, and rangeland. Runoff from farmyards and from animal confinement areas (e.g. feedlots, loafing areas) may also be a significant source of sediment. Sediment particles range in size from very fine (clays, less than 2 mm in diameter, and silts, less than 50 mm) to coarse (sand and gravel). Sediment is detached from exposed soil surfaces by rain and by surface runoff, and is transported downslope by surface runoff. Since other contaminants (nutrients, pesticides, organic matter, pathogens) are often found associated with sediment, preventing sediment detachment, reducing sediment transport, and trapping eroded sediment are critical functions of BMPs for minimizing agricultural NPS pollution.

Detachment of sediment from soil surfaces is minimized by intercepting and reducing the velocity of incident raindrops (e.g. by keeping plant residues on the soil surface) and by minimizing surface runoff (e.g. by promoting infiltration). Transport of sediment is minimized by reducing the velocity of surface runoff (e.g. by increasing surface roughness and promoting infiltration). Sediment in runoff is removed by physical trapping and by gravity sedimentation. Surface roughness provides some degree of physical trapping, as well as reducing velocity and increasing detention time available for settling of sediment.

Phosphorus

The main sources of phosphorus (P) in agricultural runoff are commercial inorganic fertilizers and organic manures and sludges applied to cropland and pastureland. Additional sources include direct runoff from animal confinement areas, crop residues, and native soil reserves. Thirty-one percent of the annual P flux to the Gulf of Mexico is estimated to come from fertilizer, 18% from manure, and 10% from municipal and industrial point sources. Forty-one percent of the annual P flux stems from sources associated with basin runoff, the most significant of which is believed to be P in sediment associated with soil erosion (Goolsby et al., 1999).

Phosphorus is found in dissolved in and particulate forms. Though the majority (55 – 85%) of total phosphorus (TP) in surface runoff is particulate P (PP) adsorbed to soil particles or organic material, only 5 - 40% of this fraction is bioavailable. Dissolved P (DP), on the other hand, is predominantly orthophosphate, which is directly bioavailable (Uusi-Kämppä et al. 1997). From 1980 to 1996, an average of 69% of the phosphorus transported to the Gulf of Mexico was in particulate and/or organic material, and 31% was transported as dissolved orthophosphate (Goolsby et al. 1999).

Particulate P consists of both organic and inorganic P adsorbed to positively charged sites on clays and hydrated iron, aluminum, or calcium oxides. There is an equilibrium relationship between sorbed and dissolved P. The equilibrium DP concentration depends on the TP concentration, the sorptive capacity of the soil, pH, ionic strength, and presence of other anions competing for sorption sites. Organic matter may be a source of competitive anions. When DP is above the equilibrium concentration value, P is sorbed onto mineral surfaces. When DP is below the equilibrium concentration value, PP is desorbed and goes into solution. At higher pH, surface hydroxides predominate over hydrates and result in negative surface charges and fewer sorption sites. As the degree of P sorption increases, the affinity for further sorption decreases due to a decrease in available sites and an increase in overall negative surface charge. Sorption of P onto accessible surface sites may be fast at first but continue over days due to diffusion of P within particles (Uusi-Kämppä et al. 1997).

Dissolved P is present in both surface runoff and infiltrating water. As water infiltrates through the soil DP is lost to available sorption sites, so phosphate is considered relatively immobile in the soil. Surface soils may have high concentrations of PP as a result. Particulate P is carried with sediment in surface runoff. The PP concentration of transported sediments is usually higher than that of the parent material due to the preferential transport of highly sorptive materials. Oxide-rich clays have a higher affinity for P than other soil particles, have a higher surface area to volume ratio than larger particles, and are easily detached and transported, so P tends to concentrate in fine sediments (Uusi-Kämppä et al. 1997).

Though much of the nutrients applied to croplands are taken up by plants and removed during harvest, excess nutrients typically remain in the soil. A nutrient balance on all US croplands in 1987 revealed that 63% of applied P remained on the land after harvest of crops and residues was accounted for (Committee on Long-Range Soil and Water Conservation Policy, National Research Council 1993). Over-application of P is especially severe when manure and sludge is land applied to meet plant N needs. Organic nutrient sources such as manure have a N:P ratio of roughly 3:1. Crops typically require a N:P ratio of 5:1 to 7:1, so manure applied to meet crop N requirements results in over-application of P. Since P is primarily adsorbed to soil particles, excess P is transported to surface waters via soil particles in surface runoff. However, dissolved P loads from pasturelands and grasslands may be higher than DP loads from croplands, especially when manure is not incorporated into the soil profile. Dissolved P loss has been observed to be directly proportional to P concentrations in surface soil (USEPA 2000a).

Pesticides

Pesticides (herbicides, insecticides, nematicides, and fungicides) are used to kill or suppress the growth of weeds, nuisance pests, fungi, bacteria, and other pathogens. Pesticides enter surface waters primarily via surface runoff, though hydrologic movement to surface water via leaching to groundwater and atmospheric deposition during application may on occasion be significant transport mechanisms. Pesticides in runoff may be dissolved in solution or sorbed to soil particles. Solubility and organic carbon partitioning coefficient (K_{oc}) are useful indicators of whether a pesticide will be found primarily in the dissolved or particulate phase. Solubility less than 10 mg/L or K_{oc} greater than 300 - 1000 mL/g indicate that the pesticide will be strongly sorbed to soil particles, especially clays and high organic matter soils (Hall et al. 1983).

Strongly sorbed pesticides are transported from fields with sediment in surface runoff and can be removed in runoff treatment areas by sedimentation and infiltration. Weakly sorbed pesticides are also transported in surface runoff and can be trapped in runoff treatment areas by infiltration and to a lesser extent by sorption. Weakly sorbed pesticides may be transported via shallow subsurface or groundwater flow as well.

Atrazine, the agricultural herbicide considered in this project, has been in use for over 40 years and is the most widely used herbicide in the U.S. Of the 76 million pounds of atrazine used in the U.S. in 1999, 83% was applied to corn, 10% to sorghum, and 3% to sugarcane. Atrazine is most frequently applied directly to soil as a pre-planting or pre-emergence weed treatment, though it may also be applied as a post-emergence foliar treatment. Effective direct soil applications require that atrazine be dissolved and incorporated into the soil mixing zone (the top 1-2 inches of soil), where it is then available for uptake by weeds (USEPA 2001b).

With a solubility of 33 mg/L and a K_{oc} of 100 mL/g, atrazine is moderately to highly mobile in soils, especially in soils with low clay or organic matter content (Wauchope, 1992). Since it does not adsorb strongly to soil particles, it is lost primarily in the dissolved phase in surface runoff. The degree of export depends strongly on storm size, slope of the soil, and the time between application and runoff: after two weeks atrazine losses in runoff were observed to be minimal (Hall et al. 1983).

Organic Carbon

Sources of organic carbon in agricultural runoff include native soil organic matter, organic manures and sludges, and plant residues. Soil organic matter contributes to improved soil physical structure (tilth), air and water infiltration, and binding of agricultural inputs such as nutrients and pesticides. Many conservation practices have the intentional or ancillary effect of increasing soil organic matter content. Conservation tillage practices, for example, minimize surface runoff and erosion by leaving organic plant residues on the soil surface, ultimately resulting in increased soil organic matter and increased water infiltration and retention capacity. At the same time, export of dissolved and particulate humic substances may potentially increase as the amount of decaying plant organic matter in the watershed increases. The combined effect of tillage practices that increase organic residues and structural practices that sequester and cycle carbon (e.g. forested riparian buffers) on organic carbon export rates in agricultural watersheds merits further study.

Pathogens

Viruses, bacteria, and protozoa found in livestock manure may be present in high concentrations in surface runoff from feedlots, pastures, and cropland or pasture fertilized with manure. Dairy farm waste, in particular that of neonatal calves, has been observed to be a significant source of *Cryptosporidium* oocysts, as discussed further in Chapter 9. Effluent from animal waste treatment systems may also be a source of pathogens. Hill and Sobsey (1998) observed that microbial indicator levels in effluent from anaerobic lagoons used to treat swine manure were on the order of 10^5 per 100 mL, much higher than the discharge standard for municipal wastewater. Indicators measured included fecal coliforms, *E. coli*, enterococci, and *C. perfringens* spores.

Pathogens may be transported from barnyards, feedlots, and manure-applied fields in surface runoff, and may be discharged directly from failing, overloaded, or unregulated waste treatment systems. Transport from waste treatment structures may be minimized through proper sizing during construction and through maintenance and inspection of critical areas such as conveyance structures and berms. Transport from barnyards and feedlots may be minimized by diverting clear runoff away from these areas and by collecting and treating contaminated runoff. Transport from fields may be minimized by conservation practices that promote infiltration and minimize erosion, by incorporation of manure into the soil, and by avoiding manure application on frozen ground or when rain is forecast. The transport of pathogens from barnyards, feedlots, and fields in surface runoff is poorly characterized. The extent to which pathogens in runoff are associated with particles, as well as the size and character of these particles, merits further study.

Removal of pathogens from surface runoff relies on settling or trapping those pathogens attached to settleable particles, and promoting infiltration of the remaining fraction. Settled or loosely bound pathogens may be subsequently reentrained by additional surface flow or flooding. Infiltration of pathogens into soil and subsequent trapping in the soil matrix is generally considered effective removal, though transport of viruses via groundwater may be a concern in some areas.

AGRICULTURAL BEST MANAGEMENT PRACTICES

Classification of BMPs

The Natural Resources Conservation Service (NRCS) codifies over 150 conservation practices in its *National Handbook of Conservation Practices* (NRCS 2002). NRCS standards define the practice, the conditions where it applies, recommendations for implementation, and operations and maintenance requirements. While many of these practices are not BMPs *per se*, the sheer number of practices illustrates the need for some system of BMP categorization.

Though the term BMP suggests pollution reduction through changes in management, in practice the term includes both structural practices, which include some physical modification of the landscape, and management practices, which rely on changes in how the land and inputs to the land are managed. A further useful distinction is commonly made between pollution prevention BMPs, which prevent contaminants from entering runoff or prevent runoff from occurring, and runoff treatment BMPs, which treat runoff after it has already occurred. While pollution prevention BMPs are typically specific to a given land use (e.g. nutrient or pesticide management for row crops), runoff treatment BMPs such as buffer strips can be used to treat runoff from a variety of land uses.

The distinction between prevention BMPs and treatment BMPs also introduces the concept of using a multiple barrier approach to controlling agricultural runoff. Pollutant loads can be reduced and reliability increased most effectively by using BMPs in sequence, starting with reduced or optimized application of chemicals or nutrients, followed by cultural practices that promote infiltration and minimize particle detachment and transport, and ending with practices or structures that capture and treat runoff.

Based on the above ideas, agricultural BMPs were grouped for the literature review, the BMP efficiency database queries, and the watershed modeling effort according to the categories in Table 2.1 below. Each of these major BMP classes may describe more than one individual NRCS Conservation Practice Standard, and each may be implemented in a wide variety of ways. Buffer strips, for example, include practices ranging from a three-zone forested riparian buffer to grassed filter strips. The list also includes four multi-practice systems: integrated pesticide management, nutrient management, grazing management, and animal waste management. These BMP systems consist of multiple individual practices designed to work together to reduce pollutant export. While it may be more difficult to characterize the effectiveness of a group of practices than a single specific practice, the BMP system approach is nonetheless believed to be useful and valid for two reasons. First, the most effective and widely recommended approach to pollutants such as pathogens, pesticides, and nutrients is not one specific practice but a suite of integrated practices. Second, performance data for some specific practices may be unavailable, because researchers often study the effects of implementing combined systems (see e.g. Meals et al. 2001).

Table 2.1
Applications of major agricultural BMPs

Application	BMP
Cropland	Conservation tillage
	Integrated pesticide management
	Nutrient management system
Pasture or range land	Grazing management system
Animal feeding operations	Animal waste management system
Structural runoff treatment	Buffer strip
	Settling basin
	Wetlands

Sources of BMP Performance Data

The amount of research conducted on the efficiency of BMPs in reducing pollutant transport to waterways has grown steadily over the past twenty years. Novotny and Olem (1994) authored a useful reference text on all types of NPS pollution and control. A useful review of NPS literature, including BMPs, is published yearly in *Water Environment Research* (see e.g. Line et al. 2000a). The EPA's *National Management Measures to Control Nonpoint Source Pollution from Agriculture* (USEPA 2000a), currently available online in draft form, provides an excellent overview of BMPs, as well as data on effectiveness and costs where available.

Findings from studies on agricultural BMP efficiency have also been compiled in at least 2 databases: the *BMP Effectiveness Database* (Dillaha et al. 2000) was developed by Virginia Polytechnic Institute for the Chesapeake Bay Program, while the *Agricultural BMP Effectiveness Database* was developed by Pennsylvania State University and others for EPA (PSU 1992). Both databases were in draft form when they were reviewed for this project, and neither has been officially released for public use. The Chesapeake Bay Program (CBP) database was selected for use on this project because of its more thorough QA/QC criteria and its higher degree of functionality. The CBP database only includes field studies conducted after 1980 that report an identifiable impact on nutrient loads or concentrations.

The CBP database was queried for studies with performance data for the selected BMPs and contaminants of concern. Query results were accepted or rejected based on the following criteria:

- Studies which do not report pollutant loading data were excluded
- Only surface water data were accepted (both subsurface and combined data were excluded)
- Studies which do not have an identifiable control were excluded
- Studies which use another BMP as a control were excluded
- Studies for which a distinction could not be made between multiple controls and treatments were excluded

Queries returned load reduction efficiencies for at least one of the contaminants of concern for over 35 different BMPs. In order to obtain BMP efficiencies with a sufficient level of generality to support the watershed modeling effort, similar BMPs were grouped together. For example, "constructed wetlands" and "wetland" were combined together under "wetland." These compiled efficiencies were then verified and adjusted as needed by consultation with experts in conservation agriculture.

While it is widely recognized that BMP efficiency varies dramatically from site to site based on many conditions (e.g. soil, topography, climate, loading, implementation variables, maintenance, etc.), it is nonetheless necessary to characterize performance in some rough fashion. Table 3.5 on page 45 summarizes typical agricultural BMP efficiencies based on median values from the CBP database, textbook values, additional literature, and best professional judgment. Practices that do not apply to certain contaminants are indicated, as are practices or contaminants for which scarce data is available. It is important to note that little data regarding the impact of BMPs on TOC export is available.

Conservation Tillage

One of the most important groups of BMPs for reducing the transport of sediment and associated pollutants from croplands is conservation tillage (often referred to as conservation tillage/residue management). These are general terms for a collection of practices intended to prevent erosion by minimizing soil disturbance and keeping plants or their residues on the soil surface. Soil erosion requires particle detachment and transport. Leaving plants and their residues on the soil prevents particle detachment by intercepting and reducing the velocity of raindrops. Particle transport is prevented by maintaining a rough soil surface that promotes infiltration. Reduced root zone disturbance increases the soil organic matter content and further prevents crusting and promotes infiltration.

Numerous specific practices are described under this category, including various types of conservation tillage, planting permanent conservation cover on retired areas, critical area planting, planting cover and green manure crops for short-term seasonal protection, and applying mulch or crop residues to protect the soil surface. Of these BMPS, conservation tillage is the most widely promoted and implemented.

The NRCS standard defines conservation tillage for areas subject to water erosion as "any tillage or planting system that maintains at least 30% of the soil surface covered by residue after planting." Major conservation tillage practices include:

- No-till / strip-till: soil and plant residues are left undisturbed from harvest through planting, and crops are seeded into a 1 – 3" wide tilled strip
- Ridge till: soil and plant residues are left undisturbed from harvest through planting, and crops are seeded into ridges that disturb roughly one-third of the soil surface
- Mulch-till: the entire soil surface is disturbed, but plant residues are exposed on 30% of the soil surface

Conservation tillage practices in many cases save time and increase profits, so adoption by farmers is quite common. From 1990 to 2000, the percent of total planted acreage under conservation tillage increased from 25% to 37%. In 2000, roughly 18% of planted acres in the U.S. were in no-till/strip-till, 18% were in mulch-till, and 1% was in ridge-till. Another 21% were under forms of reduced tillage that maintained 15 – 30% of the surface under residue cover (CTIC 1999).

Conservation tillage can be highly effective in reducing sediment export from croplands. Median TSS load reduction efficiencies from the CBP database range from 29% for ridge-till to 95% for no-till/strip-till. Novotny and Olem (1994) report sediment loss reductions of 44 – 90% for conservation tillage systems in general. Hao et al. (2001) observed, in a retrospective study of management effects on soil and SOM-C export, a roughly 90% reduction in soil and SOM-C export under no-till corn-soybean rotation compared to continuous conventional till corn.

Nutrient and pesticide loads to receiving waters typically decrease under conservation tillage. However, since these chemicals are frequently applied to the soil or residue surface, they are thus still vulnerable to transport via surface runoff. Dillaha (1990) observed that effective prevention of nutrient and pesticide transport under conservation tillage requires that nutrient and pesticide concentrations in surface soils be reduced via more effective incorporation into the soil. Thus there may at times be a tradeoff between preventing nutrient and pesticide transport by incorporating them into the soil, and preventing sediment transport by minimizing soil disturbance and maintaining plant residues. However, in warm and humid climates, increased levels of organic carbon on the surface may increase bacterial action and promote pesticide degradation. Furthermore, conservation tillage may require the use of more herbicides than conventional tillage. Despite this possibility, pesticide runoff in conservation tillage systems is typically less than under conventional tillage due to the decreased runoff and soil loss in conservation tillage systems.

Pesticide losses in nine studies (using ridge till, no-till, contour ridges, or incorporation) were 21 to 92% lower than losses under conventional tillage (USEPA 2000a). Fawcett et al.

(1994) reviewed studies assessing the effects of conservation tillage on pesticide runoff. They observed that surface runoff was the principal mechanism for transporting pesticides from fields to surface water, and that no-till practices generally reduced sediment and herbicide runoff by 90% and 70% respectively compared with conventional tillage. Pesticide losses from conservation tillage exceeded conventional tillage in cases where infiltration is limited by soil type, internal drainage, or compaction.

Many conservation tillage efficiency studies are experimental plots, the performance of which may differ substantially from field-scale practice. Clausen et al. (1996) performed one of the few paired watershed studies to assess tillage effects on pesticide loses. This four-year study on two small cornfields in central Vermont demonstrated that reduced tillage (single pass tandem disk harrow leaving 39% residue) achieved a 90% reduction in the mass export of atrazine compared to conventional moldboard plowing. Atrazine runoff was primarily in the dissolved phase, and most atrazine export occurred within 21 days of pesticide application.

Other cropland residue management BMPs include crop rotation, cover crops, green manure crops, and mulching. Crop rotation has the same principle as conservation tillage – increase infiltration by increasing soil organic matter –as well as other beneficial effects such as lower N consumption (when legumes are part of the cycle, as with the corn/soybean rotation), disrupting pest cycles, and reducing pesticide use.

Integrated Pest Management

Preventing pesticide contamination of water resources relies on a suite of mutually reinforcing control measures. The essence of Integrated Pest Management (IPM) strategies is to combine biological and cultural pest control methods with judicious use of pesticides. Fewer pesticides are used, they are only applied when economically beneficial, and they are applied in ways that minimize their transport. Steps in an IPM plan include (USEPA 2000a):

- Evaluating the cropping history of the site and past and current pest problems
- Evaluating the physical characteristics of the site to determine susceptibility to leaching and runoff
- Applying pesticides only when an economic benefit to the producer will be achieved
- Selecting the most environmentally benign pesticides, and applying them at the minimum necessary rates
- Applying pesticides efficiently and at times when runoff losses are unlikely
- Prevent spills and contamination by using anti-backflow devices and providing suitable mixing, loading, and storage areas

Important specific IPM strategies to reduce pesticide use include pest scouting, crop rotation, species and habitat diversification, proper application timing, and efficient application methods. Pest scouting can reduce application rates by adjusting the type, amount, timing, and location of applied pesticides to the specific weeds or insects at hand. Crop rotation prevents the annual buildup of pests, while diversified crops and borders are a less attractive target for pests.

Application timing and subsequent rain events have a dominant effect on the mass of pesticides exported from croplands. A light rain (or irrigation) after the application of a soil-applied pre-emergent herbicide is ideal to remove the chemical from the erodible soil surface and bring it into the target shallow subsurface zone. A heavy rain, by contrast, may prevent incorporation of the chemical into the soil and instead transport it off-site in surface runoff. Large runoff events after atrazine application on no-till systems can result in significant pesticide losses. Timing is perhaps the most critical aspect of controlling pesticide losses, yet this is the area where the farmer has the least control. Methods for minimizing the risk of pesticide loss during the first major rainfall after application include early application (when there is less likelihood of large rains), split pre- and post-emergence application (reducing the mass of pesticide at risk), and avoiding application to wet or saturated soils (Nebraska Cooperative Extension 1997a).

Numerous specific application strategies are available for conservation tillage systems. Atrazine runoff losses in mulch-till lands can be minimized by banding (spraying a band of pesticide in the crop row only, and cultivating the inter-row area); incorporation of pre-plant applications into the soil during spring tillage or seedbed preparation; or split application (using reduced pre-plant application rates followed by targeted post-emergence banding with atrazine alternatives). Foliar application rates are lower than direct soil application rates. Losses in ridge-till systems can be minimized with banding and split application. No-till systems may take advantage of early application, split application, and reduced application rates followed by spot applications where needed. (Nebraska Cooperative Extension 1997b).

No studies were located that directly assess the effectiveness of full scale, multi-faceted IPM programs on reducing atrazine losses. The variety of possible program options combined with the variability of natural physical and environmental factors makes controlled studies of this nature quite difficult. Descriptive evidence of the cumulative effect of conservation education, conservation tillage, and multi-faceted IPM strategies is provided by the agricultural communities of Osceola and Woodburn located on West Lake in south-central Iowa. Sediment and pesticide losses from increasing corn-soybean rotation acreage was reducing reservoir capacity and increasing water treatment costs. Under the Clarke County watershed management plan, a well-funded and promoted BMP implementation program provided training on soil conservation practices, soil fertility analysis, sprayer calibration, evaluation of land use, establishing realistic yield goals, assistance in implementing reduced or no-till systems, and recommendations for managing pest outbreaks. Atrazine use in the small watershed dropped by 50%, lake atrazine concentrations dropped below the 3 ppb MCL, and sediment losses dropped from 18.8 to 7.5 tons/acre (USEPA 1994a). The combined effect of reduced atrazine application rates and conservation tillage on runoff loads was not reported.

Nutrient Management

No single BMP can be expected to provide effective control of nutrient loads in runoff. Effective nutrient management requires the use of multiple practices to fine-tune the timing, quantity, and spatial distribution of manure and fertilizer application to deliver only the N and P that is needed to meet realistic crop yield goals. Nutrient management is a broadly defined practice[2] that covers the application of both manure and inorganic fertilizers to all lands. USDA (2000) outlines the principal components of a nutrient management plan:

---

[2] The NRCS Conservation Practice Standard (Code 590) defines nutrient management as managing the amount, source, placement, form and timing of the application of nutrients and soil amendments.

- Soil survey and testing: Use soil maps with aerial photos or GIS maps to determine soil productivity and identify environmentally sensitive areas. Perform soil testing at times, locations, and depths appropriate to soil types and crop needs in order to assess soil nutrient needs.
- Crop yield goals: use existing yield history data to estimate realistic yield expectations.
- Apply N and P judiciously at rates necessary to achieve realistic yield goals: To the maximum extent possible, vary manure and fertilizer application rates to meet spatially variable soil nutrient needs. 'Precision farming' systems that link GPS-based soil test data, yield history and variable rate application devices can vary N and P application spatially in order increase yield where possible and eliminate excess fertilization where unnecessary. Equipment must be calibrated to achieve accurate dose rates.
- Optimize timing, formulation, and delivery method of fertilizers and sludges to correspond with plant uptake requirements.
- Apply manure to meet P needs: Land application of sludges and manure on a disposal basis or based on plant N needs can result in high nutrient loads to receiving waters. Manure and sludge should be land applied to meet crop P needs. Excess manure should be applied to soils that are P deficient, to fields that will be planted in legumes, or to soils with low runoff potential. The NRCS Nutrient Management Standard details three methods for determining acceptable manure application rates to prevent over-application of P.
- Protect sensitive areas: Limit nutrient application in areas susceptible to erosion.
- Use runoff and erosion prevention practices to limit nutrient detachment and transport (e.g., avoid applying manure to frozen ground, incorporate manure into the soil, delay application if runoff-producing rains are forecast, use conservation tillage practices)

Since nutrient management encompasses a wide range of activities, performance can be expected to vary widely. Simple soil testing and nutrient application at agronomic rates has the potential to substantially reduce nutrient loads in some cases. A survey of over 1,900 Wisconsin farms that combined livestock and corn production showed that two-thirds of farmers applied excess N and 80% applied excess P (Shepard 2000).

The only observed efficiency data for nutrient management systems consist of measured reductions in fertilizer application rates. EPA (USEPA 2000a) reports that nutrient management plans have reduced P application rates nationwide from 0 lb/acre to over 100 lb/acre. In Maryland reductions in P application rates in 1990 averaged roughly 40 lb/acre for farms using manure, while in Pennsylvania the overall reduction in P application was 14% after the implementation of nutrient management plans. Soil testing and nutrient budgets helped farmers in the heavily farmed German Branch basin on Maryland's Eastern Shore reduce N and P application rates by 28 and 19 pounds/acre, respectively (USEPA 1994a). TP runoff load reductions for farms fertilizing with manure can be expected to be greater than those using commercial inorganic fertilizer, since the latter has the correct N:P ratio for crop needs. No studies were located that directly measured nutrient load reductions in cropland runoff as a function of changes in the cropland fertilizer loading rate. If the new application rate is roughly equal to the expected agronomic P uptake rate (a reasonable assumption since soil testing and yield expectations are implicit in the reduced application rates), then the runoff load reduction can be expected to be roughly equal to the applied load reduction.

Grazing Management

Grazing on pasture and rangelands can have negative water quality effects. Overgrazing can eliminate vegetative cover, decrease infiltration capacity, and cause erosion. Livestock (especially cattle) with free access to streams promote streambank erosion and act as point sources of nutrients and pathogens (see reviews in Line et al. 2000b, and Sovell et al. 2000). General principles for minimizing grazing impacts include lowering grazing intensity, managing the timing and duration of grazing and rest periods, distributing livestock across available land, excluding livestock from riparian and other sensitive areas, providing alternative water and shade sources, and reseeding and replanting degraded areas. Additional detail on specific grazing management system components and costs can be found in Holechek et al. (2001), USEPA (1993), and USEPA (2000a).

Livestock exclusion from sensitive areas is a common BMP for protecting source waters. Fencing cattle out of riparian areas eliminates direct access to streams, prevents streambank degradation, and allows for the development of a riparian buffer that can further reduce pollutant loads. The performance of livestock exclusion as a BMP can be expected to vary widely depending on the terrain and vegetation of both the pasture and the buffer area, the width of the buffer, and the animal stocking density. Owens et al. (1996) observed during 13 years of monitoring that exclusion reduced sediment yields from a beef cow pasture by up to 40%. Williamson et al. (1996) reported that cattle exclusion in a mixed forest/pasture watershed reduced sediment and particulate P loads to a New Zealand lake by 85% and 27% respectively. Line et al. (2000b) examined the effects of livestock exclusion on nutrient and sediment loads from a dairy cattle pasture. Observations based on 81 weeks of pre-exclusion and 137 weeks of post-exclusion data indicated that exclusion reduced TP loads by 76% and TSS loads by 82%. Meals et al. (2001) observed that a suite of livestock exclusion and riparian restoration BMPs in a northern Vermont dairy pasture achieved 28% and 49% reductions in TSS and TP export, respectively. BMPs included riparian fencing, alternative water supplies, protected stream crossings, and streambank bioengineering; BMP installation did not attempt to prevent channelized flow into the riparian area. Novotny and Olem (1994) estimate that livestock exclusion can achieve 50 – 90% reductions in TSS and TP loads.

Livestock exclusion may also significantly reduce pathogen loads, since cattle are prone to loafing and defecating directly in streams. However, few completed studies have examined this effect. Tiedemann et al. (1987) observed that exclusion could reduce fecal coliform concentrations, but that animal stocking rates had a strong effect on FC loads. Sovell et al. (2000) observed that rotational grazing resulted in lower in-stream fecal coliform concentrations than continuous grazing. Meals et al. (2001) observed that livestock exclusion and riparian restoration provided 30 – 40% reduction in concentrations of *E. coli*, fecal coliform, and fecal strep. It was also observed that bacteria levels in streams fluctuate seasonally, with low streambed concentrations in winter and spring increasing during the summer pasture period.

Animal Waste Management Systems

Numerous practices are used to minimize pollutant losses from animal feeding operations. Under current terminology these practices are lumped together as "animal waste management systems" or "comprehensive nutrient management plans" (CNMPs). CNMP components include (USDA 2000):

- Feed management: Modification of animal feed and diets to reduce nutrient concentrations in manure. Poultry and swine can be fed low phosphorus corn or enzymes such as phytase to reduce manure P levels and bring the manure N:P ratio closer to that required by plants.
- Manure handling and storage: Proper manure handling and storage, including:
  - Diversion of clean water from contact with animal waste area
  - Preventing leakage from buildings and waste conveyance and storage structures
  - Providing adequate storage for both liquid and dry manure
  - Treating manure to reduce atmospheric N loss and make it a more stable fertilizer
  - Managing dead animals to prevent contamination
- Land application of manure: Nutrient management to prevent the application of nutrients at rates that will exceed the capacity of the soil and planned crops to assimilate nutrients and prevent pollution. Application timing and methods should minimize nutrient losses.
- Land management: Conservation practices (e.g. conservation tillage, grazing management, conservation buffers) should be implemented to minimize transport of pollutants from application areas.
- Record keeping: Track soil and manure test data, manure usage and disposal, etc.
- Other utilization options: Where land application is limited, pursue other options, including sale of manure to other farmers, composting and sale to home owners, power generation, etc.

CNMP performance varies widely. Novotny and Olem (1994) report average load reduction efficiencies for improved manure handling and storage practices of 74% for bacteria, 64% for sediment, and 21% for TP. Improved farmyard runoff management was observed to substantially reduce TP concentrations in runoff at 30 south-central Florida dairy farms. Construction of drainage ditches around high intensity areas (milking barn combined with adjacent vegetation free lots) reduced average TP runoff concentrations by 62 – 87% (Havens et al. 1996).

Hill and Sobsey (1998) measured the effectiveness of primary and secondary treatment facilities on swine operations in North Carolina in reducing microbial indicators (fecal coliforms, *E. coli*, enterococci and *C. perfringens* spores). They noted that indicator levels in effluent from anaerobic lagoons (the current BMP for hog manure in the US) were on the order of 10^5 per 100 mL, much higher than the discharge standard for municipal wastewater. Anaerobic lagoons provided 1.1 – 2.2 log reduction of all indicators, except *C. perfringens*, which was reduced by 0.2 log. Of the secondary treatment systems, constructed wetlands achieved 1.1 – 2.5 log indicator reductions while a media filter and an overland flow system achieved mean indicator reductions of roughly 0.2-1.0 log. Reduction through the optimum combined system consisting of an anaerobic lagoon and constructed wetlands could therefore be expected to achieve roughly 3 – 5 log reductions for bacterial and viral indicators, and 1.5 log for *C. perfringens*.

Buffers

Buffers are one of the most widely promoted and used BMPs for treating surface runoff. This section describes different types of buffers and common characteristics; buffer strip hydrology; pollutant removal mechanisms and efficiency for different pollutants (sediment, phosphorus, pesticides, pathogens, and organic carbon); and the critical role of buffer strip maintenance in providing sustainable reductions in pollutant export.

Characteristics of Buffer Strips

Though there are many different buffer types and uses, characteristics common to most buffers include:

- Surface roughness decreases flow velocity and sediment transport capacity and promotes infiltration
- Effectiveness is highly dependent on site characteristics (slope, soils, width) and on maintaining sheet flow through the buffer
- Buffer maintenance and removal of sediment and biomass are critical to maintain long term performance
- Buffer zones may act as contaminant sources under certain conditions
- Buffer zones provide manifold benefits (e.g. biomass growth, wildlife habitat)

CTIC includes conservation buffers as one of its four primary conservation measures, and has identified ten common practices that fall under the aegis of conservation buffers (CTIC, 1999):

- Alley cropping: cultivating crops in alleys between rows of trees or shrubs
- Contour buffer strips: vegetated strips planted along contours to slow runoff and trap sediment
- Cross wind trap strips: wind-resistant plants grown perpendicular to the prevailing wind direction to prevent wind erosion
- Field border: a buffer on the edge of a field established to reduce erosion, trap pollutants, and provide turning areas for farm equipment
- Filter strip: an area of grass or other vegetation used to trap sediment and other contaminants from runoff
- Grassed waterways with vegetated filter: a vegetated channel that carries water at a nonerosive velocity to an outlet that spreads the water into a vegetated filter
- Herbaceous wind barriers: tall grass and other plants established in narrow strips across the field perpendicular to the normal wind direction
- Riparian forest buffer: an area of trees and shrubs located adjacent to surface water bodies that intercepts contaminants from surface runoff and shallow subsurface flow, enhances wildlife habitat, lowers stream water temperature, and increases streambank stability

- Vegetative barriers: parallel rows of stiff, dense vegetation planted perpendicular to the slope of the field to provide water erosion control on cropland. An alternative to terraces on sensitive soils.
- Windbreak/shelterbelt: rows of trees or shrubs that protect against wind erosion, provide shelter, etc.

Hydrology

The pollution control performance of buffer zones relies on intercepting and retaining contaminants as they pass through the buffer zone, and hence depends heavily on the hydrologic characteristics of both the drainage area and the buffer zone. Contaminants are exported from drainage areas via overland flow, shallow or deep subsurface flow, or artificial drainage (e.g. tile drains). Surface runoff entering the buffer strip infiltrates or continues as overland flow. Suspended contaminants are removed primarily by gravity sedimentation (deposition), and dissolved contaminants are removed by infiltration and subsequent adsorption or uptake.

In order to remove contaminants in subsurface flow, the flow must pass through the root zone of the buffer. While this may occur with shallow subsurface flow, groundwater may flow below the root zone and bypass treatment. Tile drains lower the phreatic surface in a field and discharge into channels and ultimately receiving waters, thus bypassing treatment in the riparian zone. Structures for the treatment of tile drain effluent (e.g. retention ponds, small treatment wetlands) may be incorporated into the riparian zone.

Sediment

Sediment removal is based on gravity sedimentation (deposition) and infiltration.[3] As surface runoff flows into buffer strips, surface roughness from vegetation and surface litter decreases the flow velocity in and immediately upstream of the buffer. This has two effects. It allows time for runoff to infiltrate into the soil, and it decreases the sediment transport capacity of the water. As the volume and velocity of flow decrease, suspended sediment settles out, with heavier particles settling at the head of the filter and lighter particles farther in. As time goes on, the head of the filter fills with sediment, surface roughness decreases, infiltration capacity decreases, and sediment is deposited further into the filter in an advancing wedge profile. As the wedge moves into the filter, the retention time in the effective settling area decreases, and removal efficiency declines.

Removal efficiency of buffer strips has been studied both in experimental vegetative filter strip plots and in long-term field studies. Findings from these studies are described below.

Vegetative filter strips. The performance of filter strips in removing sediment and nutrients from surface runoff has been studied quite extensively on experimental plots. In these studies, natural or simulated precipitation is applied to a pollutant source area (e.g. a freshly tilled field, a planted row crop, or a field spread with manure to represent pasture or feedlot conditions) and continued for a fixed time after surface runoff commences. Surface runoff enters the filter strip, and samples are collected at different points in the filter over the course of the

---

[3] Many silt and clay particles may not be removed by deposition due to unfavorable chemistry and insufficient residence time in the buffer zone. However, particles transported in infiltrating water may be removed as they pass through the soil matrix (Dillaha and Inamdar, 1997).

study. Subsurface samples are typically not collected; calculated removal efficiencies assume that infiltration provides complete retention of contaminants.

Researchers typically report removal efficiencies either in terms of percent decrease in total mass flux or percent decrease in the flow-weighted mean concentration over the course of the experiment. The latter may be more easily measured but less meaningful in some cases, since the flow rate over the strip varies temporally and spatially due to precipitation inputs and infiltration losses. The former will dilute contaminants in surface flow; the latter will concentrate them. Mass removal performance offers a better indicator of potential impact on the receiving water.

Sediment removal efficiency in vegetated filter strip studies is generally high, ranging from 75% to 95% in most cases. The median efficiency of 47 filter strip studies in the CBP database is 82%. Removal efficiency decreases as slope increases and VFS width decreases. Removal is greatest in the first few meters of the buffer. Increasing buffer strip width beyond the width of this effective removal zone does not improve removal efficiency significantly, since gravity sedimentation of the remaining fraction of suspended sediment requires more time, quiescent conditions, and favorable chemistry. Dillaha et al. (1988) observed that doubling the width of a 15' buffer increased sediment removal from 81% to 91%. Numerous researchers have observed roughly first order decay in sediment removal performance over the width of the buffer (e.g. Srivastava et al. 1998). Vegetation is observed in some cases to have a strong effect on removal performance. Stiff, dense vegetation provides greater resistance to flow and promotes deposition and infiltration (see e.g. Schmitt et al. 1999 and Lee et al. 1998). The NRCS three zone riparian forest buffer system utilizes a managed herbaceous zone to trap sediment upstream of managed and unmanaged forested zones (Dillaha and Inamdar 1997).

The most important factor affecting buffer zone sediment removal performance is the presence or absence of channelized flow through the buffer. Concentrated flow moves at a higher velocity, prevents sediment deposition, reduces infiltration, and over time creates channels that pass through the buffer. Sediment removal efficiencies in experimental plots are based on uniform sheet flow across the buffer and are thus maximum expected performances. Performance in actual settings is likely to be less, as described below in the discussion of buffer strip maintenance.

The relevance of experimental plot performance to actual applications is in some cases limited by low pollutant loading rates. Source area to buffer area ratios may be 20:1 - 40:1 in practice, compared to 2:1 – 10:1 in experimental plots, so runoff, sediment, and nutrient loading rates are often higher in practice (Dillaha et al. 1989). Finally, most VFS experiments are short term and hence do not assess removal performance over time.

Long term field studies. A limited number of studies have examined riparian processes on a larger space and time scale. Peterjohn and Correll (1984) performed overall mass balances on water, nitrogen, phosphorus, and carbon in an established riparian forest in the Atlantic Coastal Plain. The study site was a 16.3 ha subwatershed of the Rhode River drainage basin in southern Maryland, 60% of which was active fertilized cornfield. Researchers collected surface water and groundwater samples at different points along two transects running through a 50 m riparian forest buffer. Over a one-year period they found that reductions in suspended particulate matter concentrations in the first 19m of riparian forest ranged from 84% in the spring to 97% in the fall, and averaged 90% over the year. There was no significant difference in suspended particulate matter concentrations between the 19m sample point and the buffer effluent.

Cooper et al. (1987) examined long-term sediment deposition in two North Carolina Atlantic Coastal Plain riparian zones by measuring sediment accumulation above the ^{137}Cs soil horizon. There is a layer of sediment tagged with ^{137}Cs due to fallout from weapons testing in the 1950's – 1960's; this horizon can serve as a datum for measurements of sediment accumulation. Overall, 84 – 90% of the sediment removed from upland agricultural areas over a period of 20 years was retained in the watershed (including a floodplain swamp). 80% of this sediment was retained above the swamp, and over half of this amount was deposited within 100 m of the cultivated field edge. Silt and clay were observed to penetrate further into the riparian area than sand.

Daniels et al. (1996) studied the performance of grass buffers and buffers of combined grass and native ground cover in the Piedmont of North Carolina under natural rainfall conditions over a two-year period. Buffers received sheet and rill flow from forage and row crops. Removal efficiencies were skewed by one large storm event. Without this event, sediment removal ranged from 55 – 65% in the 6m grass buffers to roughly 80% in the 13m and 27m combined buffers. Including the storm, the grass buffers provided roughly 60% removal, and the combined buffers provided 30 – 60% removal. Typical sediment removal efficiencies over the 2-year period ranged from 60 – 90%.

Schwer and Clausen (1989) examined the effectiveness of a grass filter strip in treating milkhouse washwater. Washwater was applied to a 26 m long 2% slope grass filter strip twice daily over two years. Over the study period, the filter strip retained 95% of applied TSS and 89% of TP on a mass basis. Researchers also observed that 76% and 80% of TSS and TP were exported from the VFS via subsurface flow. Given this observation, it is surprising that subsurface flow is not monitored more frequently in VFS experiments.

Smith (1992) observed the importance of vegetation and the detrimental effect of channel flow in her comparison of two years of low flow and stormflow water quality data from three New Zealand pasture catchments. All were grazed with cattle and sheep, and 20% of two of the catchments was covered by 25 – 35 m wide riparian buffers of mature pine forest. Sediment concentrations were 52-62% less in the pastured watershed than the afforested watershed under low flow conditions, and 76 – 89% less under stormflow conditions. The increased sediment yield from the afforested catchments was attributed to shading by the pine canopy, which decreased close ground cover and resulted in concentrated flow.

Phosphorus

Removal of P in buffer zones relies on sedimentation, infiltration, and plant uptake. Particulate P is removed with sediment. PP removal efficiencies are typically lower than those observed for bulk sediment, however, due to the concentration of P ('enrichment') in the smaller, slower settling sediment fraction. Removal can be expected to diminish as flow velocity increases and residence time decreases. High flows may flush sediment and sorbed P from the buffer zone.

Removal of DP is based on sorption onto available sites. Sorption and desorption depend partly on the equilibrium DP concentration at that point in the system. Since sorption may be slow, removal of DP is increased by promoting infiltration, both in the field and in the buffer strip. Plants and microbes assimilate inorganic P and convert it into organic P in biomass. The significance of this pathway may vary widely: Schwer and Clausen (1989) observed that of the 89% of TP retained in the buffer, only 2.5% was assimilated into biomass. The balance was presumably sorbed onto soil, the capacity of which could be expected to decline over time. Peterjohn and Correll (1984) calculated that 17% of the TP uptake by trees was stored in woody biomass, the remainder being stored and shed in leaves.

TP and DP removal efficiencies vary widely. Peterjohn and Correll (1984) observed 81% retention of TP in the Rhode River watershed, while Lowrance et al. (1984) observed 24% retention in the Little River (Georgia) coastal plain watershed. Uusi-Kämppä et al. (1997) suggest that the maximum practical TP removal is around 50%.

Removal of dissolved P (predominantly ortho-P) from surface and subsurface flow is generally considered substantially less than PP removal (Lowrance et al. 1997). In some cases buffer zones release DP (e.g. Osborne and Kovacic 1993). Data suggest that P uptake during the planting and growing season is followed by P release during the dormant season, perhaps because of leaf decay. In general, biomass harvest removes assimilated P from the buffer zone, and biomass decay releases organic DP (Uusi-Kämppä et al. 1997). Long term P control thus depends on biomass and sediment management.

Pesticides

Buffer strips may provide effective removal of pesticides. Degradation is provided by biological and nonbiological processes, including microbial detoxification, plant absorption and metabolism, chemical catalysis, photodecomposition, and volatilization. The rate of degradation varies between pesticides and with environmental conditions (e.g. temperature, pH, humidity, UV exposure, organic carbon concentration, microbial activity).

Atrazine is retained in buffers primarily by infiltration (see e.g. Arora et al. 1996). Removal efficiencies vary widely, due in large part to the timing and high variability of natural rainfall events and the extent of infiltration in the buffer strip. Large rainfall events soon after herbicide application transport high loads of atrazine into the buffer zone. If the surface flow exceeds the infiltration capacity of the buffer, minimal treatment is achieved. Conversely, buffers that provide near complete infiltration of runoff will provide near complete atrazine removal. The median atrazine load reduction of 8 filter strip studies in the CBP database is 98%. Other studies have shown typical removals in the 30 – 70% range.

Pathogens

No studies were located that directly assess the ability of riparian buffers to remove *Cryptosporidium* oocysts. A limited number of studies have assessed the ability of buffers to remove indicator bacteria (total coliforms, fecal coliforms, or fecal streptococci) from surface runoff. The median bacteria load reduction of 3 filter strip studies in the CBP database is 55%.

Removal of indicator bacteria and other pathogens in buffer strips is highly variable. Effective removal requires a high degree of infiltration. Large storms, channelized flow into or through the buffer strip, and reduced buffer strip infiltration capacity can significantly reduce pathogen removal performance. Dissolution, reentrainment, and flushing of settled particle-bound pathogens also decrease performance. Export of pathogens in subsurface flow has received insufficient attention and may further reduce performance, especially in soils with significant preferential flow paths. Finally, the relevance of coliform studies to the transport and fate of environmentally resistant pathogens is not clear.

Organic Carbon

Riparian buffers have been observed to increase the soil organic matter fraction in the riparian zone. Marquez et al. (1999) observed an 8.5% increase in soil organic matter carbon (SOM C) over 7 years of buffer growth from cultivated land to cool season grass or poplar/ switchgrass multispecies riparian buffers. Since SOM correlates with soil productivity and pollutant removal capacity, sorption of highly sorbed pesticides may thus improve over time in some buffer systems. It is not known if this increase in SOM C will lead to an increase in C export to adjacent streams.

A limited number of studies have addressed carbon fluxes through riparian buffer zones. Jordan et al. (1993) examined groundwater in a transect through a riparian forest adjacent to cornfields and found that groundwater dissolved organic carbon (DOC) concentration remained constant throughout the buffer. Peterjohn and Correll (1984) observed an annual average of 60% and 64% reduction in surface runoff particulate organic carbon (POC) concentration in 19 and 50 m of riparian forest, respectively. Significantly, surface runoff DOC concentration increased by a factor of 2.75 through the 50 m buffer. Srivastava et al. (1996) observed 11 – 30% mass reductions of TOC through 6 – 18 meter fescue buffers receiving poultry litter runoff. Riparian zones can thus act as sinks for POC, but also as sources of DOC. Further investigation into carbon fluxes through riparian zones is warranted.

Maintenance

Maintenance of buffer zones is essential for long-term performance. Preventing channelized flow through the buffer is a critical factor affecting buffer zone pollutant removal performance. Concentrated flow moves at a higher velocity, prevents deposition of particulate contaminants, reduces infiltration, and over time creates channels that pass through the buffer. In one of the few experiments to consider concentrated flow, Dillaha et al. (1988) observed a 31 - 58% decrease in sediment removal performance in plots receiving concentrated flow generated by a 4% cross-slope.

Dillaha et al. (1989) later investigated the effectiveness of filter strips under actual field conditions. In a qualitative assessment of 27 Virginia farms with grass buffer strips, researchers estimated that 60% of runoff entered the strips as channel flow, not uniform sheet flow. This was most prevalent in large fields with internal drainage features in hilly areas. It was also promoted by sediment accumulation at the head of buffer areas and by berms at the edge of buffers created by moldboard plowing. In both these cases runoff flows parallel to buffers until it reaches a low point, where it enters the buffer as channelized flow. Within the buffer, infrequent mowing was observed to promote weed growth, reduction in close ground cover, and reduced sediment and nutrient trapping efficiency.

Channel flow through buffer zones can be prevented by uniform grading at the head of the buffer; by water bars, which redirect water flowing parallel to the buffer back into the buffer; and by level-lip spreaders, ditches which discharge evenly into the buffer. Channels that form in the fields and pass through the buffer zone may be diverted into level-lip spreaders or may be treated separately with small wetlands.

Sediment removal is also critical to prevent the buffer zone from filling with trapped sediment and become ineffective at removing further loads. Depending on the relative elevations of the receiving channel and the buffer zone, flooding may inundate the buffer zone periodically and either carry away trapped sediment or deposit additional sediment. Large storm events may flush accumulated sediment into receiving waters. Periodic sediment regrading or removal is thus essential for maintaining removal performance.

Settling Basins

NRCS distinguishes several different types of runoff treatment basins (e.g. water and sediment control basins, settling basins, detention ponds, wet detention pond, dry detention pond). All remove sediment and sediment-associated contaminants via gravity sedimentation. Performance will vary with size, detention time, and loading rate. Novotny and Olem (1994) report load reduction efficiencies of 40 to 90% for sediment and 40% for TP. The median load reduction for settling basin in the CBP database is 62% for sediment (9 studies) and 30% for TP (4 studies). Additional research is needed to establish performance data for TOC, *Cryptosporidium*, and atrazine.

Wetlands

Wetland performance is poorly characterized in the agricultural NPS/BMP literature. The median sediment load reduction of 3 wetland studies in the CBP database is 67%. Mean removal efficiencies of 70% for sediment and 30% for TP are averages for surface and subsurface flow systems based on the North American Wetland Treatment System Database, as reported in Kadlec and Knight (1996). Additional studies are needed to establish performance data for TOC, *Cryptosporidium*, and atrazine.

CHAPTER 3
METHODS: WATERSHED MODELING

INTRODUCTION

This chapter consists of three main sections. The first section is a literature review that provides an overview of different watershed models and guidance in selection of watershed models. The second section provides a more detailed description of the watershed model selected for use in this project (the Watershed Treatment Model), including model default values for pollutant export coefficients and urban and agricultural BMP efficiencies. The third section provides details on the modeling scenarios developed for this project.

LITERATURE REVIEW: WATERSHED MODELS

Though the discipline of watershed modeling has advanced considerably over the past thirty years, it is by no means mature today. Continued advancement is promoted by the increasing awareness of the impact of land use and nonpoint source pollution on water quality, as well as by regulatory drivers such as Total Maximum Daily Loads (TMDLs) and Source Water Assessment Programs (SWAPs). The high degree of activity in the watershed modeling field is reflected in the wide variety of currently available models, the continued refinement of existing models and development of new models, and the increasing availability of data, much of it in GIS format. As watershed management becomes a real activity in more milieus, and as the monetary value of land-use trade-offs grows, practical and scientifically defensible watershed modeling tools become increasingly important as planning, management, and negotiating tools.

Numerous watershed loading models have been developed. This review does not attempt a comprehensive description of all models, and provides instead an introduction to important characteristics of watershed modeling and a description of selected watershed models, their assumptions, requirements, and output.

Model Characteristics

Watershed loading models are distinct from receiving water models. A watershed loading model provides an estimate of the mass or concentration of a pollutant discharged from point and nonpoint sources within a watershed to a water body. A receiving water model describes the transport and fate of the pollutant after it enters the water body. Conceptually, watershed models must simulate two processes. First, they must account for the hydrologic transformation of rainfall to runoff. Second, they must account for processes that affect water quality. These accounting methods may attempt to simulate fundamental processes, or they may rely instead on empirical correlations and surrogate parameters.

Hydrologic Accounting

Hydrologic processes affecting surface and subsurface runoff include precipitation, evapotranspiration, infiltration, interception, and storage. Precipitation (or snowmelt) generates rainfall excess after surface storage (from interception and depression) and infiltration have been satisfied. Evapotranspiration has a less immediate effect on surface runoff, but determines in large part the soil antecedent moisture conditions, which in turn affect infiltration (Novotny and Olem 1994). Though numerous equations based on measurable properties have been developed for quantifying these processes, data requirements may easily exceed available data, and accuracy is often low due to spatial variability of properties.

Because of the limitations of explicitly calculating infiltration, surface storage, and evapotranspiration, the Soil Conservation Service (SCS) Curve Number (CN) method has found widespread usage. This method calculates rainfall excess as a function of precipitation and tabulated CN values. CN values depend on land use and cover, soil hydrologic groups, and antecedent soil moisture conditions. More complex models may provide a more detailed simulation of hydrologic processes, including unsaturated and saturated subsurface flow (Novotny and Olem 1994).

Rainfall excess is water available for surface runoff. Common empirical methods for converting rainfall excess to surface runoff include the Rational Method, unit hydrographs, and synthetic unit hydrographs (Novotny and Olem 1994).

Pollutant Accounting

Watershed loading models must combine a description of hydrologic processes with some approximation of pollutant generation, transport, and transformation processes. Common pollutants of concern include sediment, nitrogen, phosphorus, metals, pesticides, and BOD/COD. The number of pollutants and processes simulated varies depending on the complexity of the model, and varies from empirical estimations of sediment transport to an explicit accounting of nutrient speciation and adsorption. Sediment losses during transport via overland flow are often accounted for with a delivery ratio, which is the ratio of the pollutant load on the receiving water to the pollutant mass available for export. Adsorbed pollutants may be tracked with sediment by using potency factors, which relate pollutant and sediment concentrations.

Spatial Scale

The goals of the watershed modeling task, and the tools appropriate to achieve those goals, will depend on the size of the area of interest, which may range from basin-scale (e.g. the Chesapeake Bay watershed) to field-scale.

Watershed loading models are typically classified as lumped or distributed parameter models. Lumped parameter models treat the catchment, or large areas within the catchment, as one unit. Salient catchment characteristics are combined together in a simplified empirical equation that transforms model input to model output. Hydrologic processes and pollutant transport and transformation processes may not be modeled explicitly, and model parameters may have no physical meaning. Determination of parameter values may require calibration with flow and pollutant data sets.

Distributed parameter models divide the watershed into smaller areal units (finite elements), each with relatively homogeneous properties. Spatial discretization may be based on sub-catchments or may result from a super-imposed grid or mesh. Physical characteristics such as rainfall, soil characteristics, land cover, and pollutant export coefficients are averaged across the unit. Each element is linked with adjoining elements, and model output is generated by simultaneously solving mass-balance equations for each cell. Data requirements for distributed parameter models are extensive, so distributed parameter models are frequently integrated with GIS software. The increased availability of powerful personal computers and GIS data and software has facilitated the use of distributed parameter models. However, limitations in monitoring data (e.g. inability to account for the spatial variability of rainfall) may still restrict the accuracy of model predictions.

The level of spatial discretization imposed on the watershed has been observed to have a significant impact on model predictions. Terrain, soils, and land-cover data is often available at a finer discretization than is desired in the model, so some degree of catchment and parameter aggregation is required. FitzHugh and Mackay (2000) observed that sediment yields predicted by the SWAT model (described below) dropped by 44% from the coarsest to the finest watershed delineation. Sensitivity to spatial discretization was also observed in the ANSWERS model by Brown, Brian, and Walsh (1993) and in the AGNPS model by Vieux and Needham (1993).

Temporal Scale

Watershed simulations are typically performed on an event, annual, or continuous basis. Event models simulate the response of a watershed to a storm of specified size. Event models are useful because peak pollutant loads are event-driven, data requirements are minimal, and computer run-time is short. However, since runoff depends on both storm size and antecedent moisture conditions, event simulations assume that rainfall and runoff have the same return period (Novotny and Olem, 1994).

Annual simulation models estimate pollutant loads on a yearly basis. Though these models offer no predictive ability for specific storm events, they do provide a good measure of watershed health over time. Land-use pollutant export coefficients are often measured or reported on an annual basis.

Continuous simulation models simulate the watershed response to changing meteorological conditions over time. Hydrologic and pollutant transport/transformation processes are simulated during each time step. Run-time is greater than event or annual models. Data requirements are extensive, and the temporal discretization of the available data may be coarser than the desired model time step.

Time-step selection may have a significant impact on model results. Rudra, Dickinson, and von Euw (1993), for example, observed that model output (runoff, soil loss and sediment yield, and soil calibration parameters) was extremely sensitive to small variations in the rainfall data time step for both a field-scale continuous simulation model and an event-based watershed-scale model.

Model classifications

Several different schemes have been advanced for classifying watershed loading models. Novotny and Olem (1994), for example, categorize watershed models as simple statistical routines and screening models, deterministic hydrologic models, and stochastic models. EPA (USEPA 1997a) classifies models as simple, mid-range, and detailed. Though the EPA classifications are somewhat subjective, they do offer a quick indication of model complexity

Simple or screening-level models use empirically developed relationships and tabulated parameters to provide a rough estimate of pollutant loads. Though the level of accuracy of these models is typically low, they are easy to implement and can provide a preliminary identification of potential significant sources of NPS within a watershed. They can also be used for qualitative comparisons of alternative management scenarios. Use of screening models requires only a calculator or spreadsheet software, and data requirements are minimal. The dominant input variable is land-use type. Other input variables may include slope and soil characteristics. Output is typically on annual load basis.

Mid-range models maintain much of the empiricism of screening models, but accept more geographic refinement and may offer continuous or event simulation as well. They can be used to identify important pollutant source areas and target sources best suited for BMP implementation.

Detailed models offer a more mechanistic description of hydrologic and pollutant transport and fate processes. They can be used to predict flow and pollutant concentrations at locations throughout the watershed over time. Detail and predictive power increase, but skill level and data requirements are extensive.

Increased model complexity does not guarantee increased accuracy and predictive ability. Models are in all cases simplified representations of real, complex phenomena. Predictive ability depends not only on our ability to identify and mathematically describe relevant physical processes, but also on the quality and availability of data, the size of the model domain, the degree of temporal and spatial discretization we impose on the model domain, and the parameters of concern. Novotny and Olem (1994) rank the following parameters in order of increasing possibility of error:

1. Hydrology (impervious surface)
2. Hydrology (pervious surface)
3. Sediment
4. Phosphates and metals
5. Nitrogen and organic chemicals
6. Bacteria

Model selection criteria

Model selection should be based initially on identifying models that meet the objectives of the modeling effort, which may include:

- Assess the magnitude of point vs. nonpoint source pollutant loads
- Locate significant sources of the pollutants of concern
- Predict the reductions in pollutant load possible through the implementation of BMPs
- Prioritize pollutant source areas that should be targeted for BMPs
- Design BMPs to achieve specified reductions in pollutant loads

Other criteria include the model's ability to describe the spatial extent of the area of interest under the desired temporal conditions (e.g. continuous vs. event). Model sophistication and accuracy of predictions must be weighed against data requirements and availability, as well as the skill level required for effective use. In general, simplicity is an overriding principle. The selected model should only be as complex as necessary in order to achieve the stated objectives.

Models that are too complex are likely to waste time and resources, and may yield results that are inaccurate or more precise than necessary. Finally, it is important to choose a model that is well documented and tested, and that has an experienced community of users.

The number of watershed loading models has grown significantly from the 1970s to the present, and numerous reviews are available (see e.g. Giorgini and Zingales 1986, Dillaha 1990, USEPA 1991, Novotny and Olem 1994, Jolankai et al. 1999). Perhaps the most useful, comprehensive, and readily available is the EPA's 1997 publication *Compendium of Tools for Watershed Assessment and TMDL Development* (USEPA 1997a). This free publication presents uniform descriptions of major watershed loading models, receiving water models, and ecological assessment techniques, as well as guidance on model selection, calibration, and verification. Model descriptions are accompanied by usage examples and thorough documentation. The interested reader is referred to this publication for additional guidance on watershed model selection and usage.

Screening Models

Universal Soil Loss Equation (USLE)

The most widely used method for predicting soil loss from sheet and rill erosion is the Universal Soil Loss Equation (USLE) (Novotny and Olem 1994). The USLE or its later versions MUSLE (Modified USLE) and RUSLE (Revised USLE) have been incorporated into numerous watershed loading models, including the EPA Screening Procedures, SLOSS-PHOSPH, Watershed, GWLF, AGNPS, STORM, SWAT, and SWMM (USEPA 1997a).

The USLE does not explicitly represent mechanisms of particle detachment, transport, and deposition. The USLE is instead an index method that calculates the soil loss rate from tabulated factor values that represent the impact of climate, topography, soil, and land use on sediment detachment by rainfall and overland flow. Developed by Wischmeier and Smith in 1965, the USLE predicts areal soil loss according to the equation and variables described below (Wischmeier, 1976):

$$A = RKLSCP$$

A (average annual soil loss). The USLE returns the value of A, the average soil loss for the time interval of R (usually per year), in units of factor K (usually tons per acre).

R (rainfall erosivity index). R, the rainfall erosivity index, is a measure of the erosive forces of rainfall and runoff, and is equal to the sum of the erosion indices (EI) of each storm in the time period. R accounts for the interaction between storm size and intensity. Soil losses have been observed to be proportional to the product of the total kinetic energy of the storm (E) and its 30-minute intensity (I). Regional annual average R values are available from iso-erodent maps or tables, though R values can vary significantly over seasons (Novotny and Olem, 1994).

K (soil erodibility factor). K, the soil erodibility factor, is an empirical measure of the erosion potential of a soil, and has units of mass/area. K can be measured directly on unit plots under standard conditions, or calculated from grain size distributions with the use of nomographs or tables (Novotny and Olem, 1994). K is an empirical index of erodibility. It is not a soil property itself, but is influenced by soil properties such as texture, permeability, detachability, and transportability.

The product RK in the USLE is an estimate of soil loss from experimental unit plots of specified dimension, steepness, cover, and treatment. R and K have units, whereas the other RUSLE factors are dimensionless.

LS (slope length and steepness factor). LS is a dimensionless parameter that accounts for the effects of slope length, steepness, and shape on sediment production. Novotny and Olem (1994) provide further detail for calculating LS. The slope length L is defined as the distance from the origin of overland flow along the flow path to the point where deposition begins to occur on concave slopes or to a concentrated flow channel. Slope steepness, shape, and position along the slope have a large impact on erosion.

C (cover-management factor). C, the vegetative cover and cropping management factor, is the most direct measure of the impact of management practices on erosivity. C is a dimensionless factor that accounts for the effects of canopy, ground cover, surface roughness, time since last disturbance, amount of roots in the soil, and soil organic matter content. C values change throughout the year, so the C factor is an average annual value weighted according to the variation of rainfall erosivity over the year.

P (supporting practices factor). P, the conservation practice factor, is a dimensionless factor used to account for reduction in erosion due to best management practices (e.g. contouring, strip cropping, terraces, sediment basins, grass hedges, silt fences, straw bales). These BMPs are applied in addition to the basic cultural practices represented by the C factor. Deposition on concave slopes is also accounted for with the P factor.

Use of the USLE. Wischmeier (1976) caution practitioners to stay within the limits of the original USLE parameters. The USLE returns the amount of soil lost or detached from a plot of land, so the sum of soil loss estimates from a number of plots in a drainage area measures the amount of soil moved from its original position. This is not the same as the effective sediment yield for the area (i.e. the sediment load on the receiving water), since sediment detached from one area may be deposited in another at a field border, depression, or the bottom of a slope. Sediment yields are estimated by combining sediment losses with estimates of sediment deposition (e.g. delivery ratio) and sediment addition. The RUSLE facilitates computation of sediment deposition (see below).

Wischmeier (1976) also advise caution when using the USLE on a large scale. It would be incorrect to apply the USLE to a large watershed by using overall averages of L, S, K and C. Factor values are inter-related, so correct use of the equation relies on associating the factors with each other according to actual field conditions. Finally, the developers note that while the USLE can be used to predict soil loss from a given storm event, accuracy is expected to be much lower. Tabulated factor values produce relatively accurate results on a long-term average basis, however predictive power for a given storm is much lower due to inherent temporal fluctuations in the relationships between the factor values.

RUSLE. The USLE has been continuously updated and revised since its introduction (see e.g. Renard et al. 1991). The most current version of RUSLE (Version 1.06) and data files can be downloaded at no cost from the USDA Agricultural Research Service National Sedimentation Laboratory. The current version of RUSLE allows for computation of sediment deposition in concave slopes, channels, and basins as a function of sediment characteristics, as well as improved methods for computing the slope length factor and the effectiveness of ground cover.

RUSLE is currently undergoing another major revision, RUSLE2, which will be equipped with a graphical user interface and will be easier to learn and use. Substantive changes include the temporal integration of the erosivity distribution and time varying K, C, and P factors, as well as improvements to the residue decomposition, roughness, runoff, and hydraulic computations.

Urban Buildup / Washoff

Urban nonpoint source loads are frequently modeled using a buildup-washoff model. This concept is used in simple screening models as well as the more advanced STORM, DR3M-QUAL, and SWMM models. The buildup component is based on researchers' observations that urban curbside pollutant concentrations build up over time until equilibrium is reached between deposition and loss to pervious and hydrologically inactive surfaces. Washoff occurs when the accumulated pollutants are flushed from the surface during rainfall. Numerous build-up and washoff functions and parameter values have been presented (see e.g. Barbe, Cruise, and Mo 1996). Zhang and Yamada (1996), for example, observed that although the buildup and washoff functions are usually calibrated separately, field samples usually represent a combination of both processes. They combine the buildup and washoff equations into one model, and use optimization techniques to select appropriate parameter values. Novotny and Olem (1994) review urban nonpoint source load estimation techniques and data sources.

Export coefficients

Export coefficients or unit area loads (e.g. lb/ac/yr or kg/ha/yr) can be used to estimate annual nonpoint source pollutant loads as a function of major determining parameters such as land use, land cover, and soil type. Tabulated values from numerous studies are available; see e.g. USEPA (1976) and Beaulac and Reckhow (1982). Mattikalli and Richards (1996) and Johnes (1996) exemplify the use of export coefficients and detailed land-use data to assess the impact of land-use changes on water quality.

Simple Method

The Simple Method (Schueler 1987) is a hand-calculated screening procedure for predicting annual pollutant loads and potential load reductions in small urban watersheds under alternative development scenarios. Pollutant loads are predicted as a function of drainage area, impervious cover percentage, annual rainfall, and typical urban runoff pollutant concentrations. Pollutants modeled include phosphorus, nitrogen, chemical oxygen demand (COD), biological oxygen demand (BOD), and metals (USEPA 1997a).

Watershed Treatment Model

The Watershed Treatment Model, Version 3.0 (Caraco 2001) is a screening-level, spreadsheet-based model that predicts annual pollutant loads delivered to water bodies. The model predicts loads from primary sources, calculated as a function of land area, as well as secondary sources such as CSOs, SSOs, and channel erosion. Loads from pervious surfaces are calculated using export coefficients, while impervious surface loads use the Simple Method. Pollutants modeled include sediment, nitrogen, phosphorus, organic carbon, fecal coliforms, and pesticides. More detailed descriptions of model algorithms and assumptions are provided below and in Caraco (2001).

STEPL

The Spreadsheet Tool for the Estimation of Pollutant Loads (STEPL Version 2.0) is a simple spreadsheet based model for calculating nutrient and sediment loads and load reductions from BMPs (Tetra Tech 2001). It computes nutrient and sediment loads and load reductions due to BMPs for cropland, pastureland, farm animals, feedlots, urban runoff, and failing septic systems. Runoff is calculated using the NRCS curve number method, while sediment loads are calculated using the USLE and delivery ratios. Built in datasets and BMP efficiencies allow easy use and quick calculations. STEPL is expected to be made available by USEPA.

Other Methods

Other screening methods include the EPA Screening Procedures (USEPA 1985), USGS Regression Method, the Federal Highway Administration Model (FHWA), Watershed Management Model (WMM), and SLOSS-PHOSPH. Spreadsheet software can also be used to facilitate calculation of NPS pollutant loads derived from land use and cover statistics and pollutant source equations (e.g. USLE, buildup-washoff, or unit area loads). Walker, Pickard, and Sonzogni (1989), for example, use a spreadsheet model based on the USLE and export coefficients to estimate nonpoint source phosphorus loads on a Wisconsin Lake.

Mid-Range Models

AGNPS

The Agricultural Nonpoint Source Pollution Model (AGNPS) is a distributed parameter, GIS-based model developed and maintained by the USDA Natural Resources Conservation Service (NRCS) and Agricultural Research Service (ARS) (Young et al. 1989). The model uses a square grid system to predict point and nonpoint source pollutant loadings (sediment by particle size class, nutrients, and pesticides) within agricultural watersheds, and to assess BMP effects. Hydrology is based on the SCS curve number; sediment loads from erosion are based on the RUSLE. Though historically limited to event simulation, the latest version of the model (AnnAGNPS) has been updated to handle continuous simulation.

GWLF

The Generalized Watershed Loading Functions (GWLF) were developed by Cornell researchers to estimate point and nonpoint source nutrient and sediment loads from urban and agricultural watersheds and to evaluate BMP effectiveness (Haith and Shoemaker, 1987). This continuous simulation model generates runoff with the SCS curve number, sediment loss with delivery ratios and the MUSLE, and urban nutrient loads with a buildup/washoff model. The model uses tabulated and default parameters extensively and in principle does not require calibration (USEPA 1997a). The model uses a daily time step and can simulate annual and seasonal loads. Recent versions incorporate a septic system component and graphical output aids. The GWLF model has been incorporated in the EPA's Watershed Screening Model (WSM), where it profits from an enhanced GUI as well as links to national databases.

SLAMM

The Source Loading and Management Model (SLAMM) is a continuous simulation, planning level, urban NPS modeling tool designed to identify pollutant sources and evaluate the effectiveness of source area and outfall control practices (e.g. infiltration practices, wet detention ponds, porous pavement, street cleaning, catch basin cleaning, grass swales) (Pitt, 1993). Many urban runoff models emphasize large storm events, however stormwater quality problems are often associated with smaller storm events. SLAMM hydrology and water quality algorithms are selected for small event simulation.

Detailed Models

HSPF

Hydrological Simulation Program - FORTRAN (HSPF) is a multi-function program that allows continuous and storm-event simulation of watershed and water body hydrology, point and nonpoint source contaminant loading, and receiving water quality. HSPF simulates up to seven pollutants (sand, silt, clay, BOD, orthophosphate, ammonia or nitrate, and one user-specified organic pollutant), and numerous soil processes (e.g. leaching, sorption, soil attenuation, soil nutrient transformations). The receiving water component models in-stream hydraulic, water temperature, sediment transport, nutrient, and sediment-chemical interactions (USEPA 1997b).

Developed from the Stanford Watershed Model (Crawford and Lindsey 1976), HSPF is a robust and widely used NPS modeling program. The Chesapeake Bay Program, for example, uses an HSPF model of the Bay watershed to simulate nutrient and sediment loads on the Bay. USEPA, USGS, and the Army Corps of Engineers are committed to using HSPF (Whittemore and Beebe 2000).

HSPF is limited in several respects, however. It is a complex tool that requires significant user expertise for meaningful use, so accessibility is restricted. Calibration and pre- and post-processor packages from USGS, as well as integration into the BASINS package (see below), have added graphical user interface (GUI) capacity and increased ease of use. Though they facilitate access, these improvements may also serve to obscure model assumptions (Whittemore 2000). Data requirements for simulation, calibration, and verification are extensive. Finally, HSPF is a lumped parameter model that requires calibration, so its ability to evaluate the impact of BMP implementation is limited. Since many model parameters are not measurable quantities, calibration is essential for every use. And, since the model is calibrated to existing conditions, it is difficult to modify parameters for future conditions (Novotny and Chesters, 1981).

SWAT

Soil and Water Assessment Tool (SWAT) is a USDA-supported, long-term, basin-scale, continuous simulation model developed to assess the impact of management practices on NPS pollution and water supplies (Arnold et al. 1998). Predecessor models include SWRRBWQ,

GLEAMS, and CREAMS. Like HSPF, SWAT models the watershed with a network of linked upstream and downstream watershed elements. In SWAT the model elements are component subbasins. Hydrologic processes modeled include precipitation, surface runoff, percolation, lateral subsurface flow, groundwater flow, evapotranspiration, snowmelt, and transmission losses. Unlike HSPF, SWAT can in principle be used without calibration, since model parameters are measurable or tabulated values. The SCS Curve Number method is used to simulate surface runoff, and the MUSLE is used for erosion calculations. SWAT is supported by the USDA Agricultural Research Service, Grassland, Soil and Water Research Laboratory.

STORM

Storage, Treatment, Overflow Runoff Model (STORM) is a continuous simulation stormwater runoff model developed by the Army Corps of Engineers (HEC 1977). Hydrology is based on the SCS curve number method, and erosion is calculated from the USLE. Other water quality parameters are simulated by a buildup-washoff model. STORM has been most widely used for assessing treatment and storage options for combined sewer overflows (CSOs) (USEPA 1997a).

SWMM

The Storm Water Management Model (SWMM) was originally developed for urban stormwater management, and has been widely used for detailed simulation of urban hydrology and hydraulics, including combined sewer overflows (USEPA 1988). It can track up to ten pollutants, and has been updated to handle continuous simulation and more complex watersheds and land uses. Data requirements for calibration/validation are extensive. A GUI and post-processor are available (USEPA 1997a).

ANSWERS

The Areal Nonpoint Source Watershed Environment Response Simulation Model (ANSWERS) is an event-based, distributed parameter model developed to evaluate the effects of land use and management practices on water quality. Preparation of data input files is complex, though more flexibility is available in output format (USEPA 1997a). Borzoi and Dillaha (1996) describe the performance of a continuous simulation version, ANSWERS-2000, developed to simulate long-term average annual runoff and sediment yield from agricultural watersheds without calibration.

BASINS

Better Assessment Science Integrating Point and Nonpoint Sources (BASINS) is a comprehensive watershed assessment tool that integrates modeling tools with water quality and pollutant source data on a GIS platform (USEPA 1998). Key modeling tools in the package include two watershed loading and transport models (HSPF and SWAT), a simple nonpoint source annual loading model (PLOAD), and a receiving water model (QUAL2E). Integrated databases and data mining tools facilitate access to spatial, environmental monitoring, and point source discharge data. The BASINS graphical interfaces, linked programs, and GIS software greatly simplify model usage.

BASINS was developed in part to aid the process of developing TMDLs for impaired water bodies, and as such has received widespread usage, as well as continued support from EPA, including release of Version 3.0 in 2001. Whittemore and Beebe (2000) summarize expected model changes, and express concern regarding database integrity, insufficient peer review, and lack of available training.

MODEL DESCRIPTION: WATERSHED TREATMENT MODEL

Watershed modeling for this project was performed using a modified version of the Watershed Treatment Model, Version 3.0 (Caraco 2001a). The Watershed Treatment Model (Caraco 2001b) is a screening-level, spreadsheet-based model that predicts annual pollutant loads delivered to water bodies. The model predicts loads from nonpoint sources, which are calculated as a function of land area, and from secondary sources such as point sources, channel erosion, and CSOs/SSOs, which are not calculated as a strict function of land area. Detailed descriptions of model algorithms, assumptions, and supporting data for selection of model default values is provided in Caraco (2001b). Model documentation in this report is limited to a brief description of the basic WTM model methodology and the underlying model assumptions. Additional description of assumptions for individual modeling scenarios is also provided.

Primary Sources

The WTM divides primary sources into two classes: urban land uses, or those characterized by a high degree of impervious cover, and rural land uses, with little or no impervious cover. Urban land use classes include low, medium, and high-density residential housing, commercial, roadway, and industrial. Urban loads are calculated using the Simple Method, which predicts stormflow pollutant loads as a function of drainage area, impervious cover percentage, annual rainfall, and typical urban runoff pollutant concentrations (Schueler 1987). Default values for percent impervious cover, urban runoff concentrations, and calculated pollutant export coefficients assuming 35" annual rainfall are summarized in Table 3.1.

Rural or pervious areas include forest, cropland, and pasture/range. Pollutant loads from these areas are calculated as a function of land area and annual pollutant export coefficients. Default pollutant export coefficients are selected from the literature to represent national medianvalues (Caraco 2001a) and summarized below.

Table 3.1
Default pollutant export coefficients used in Watershed Treatment Model

Cover type (calculation method)	Land use	Impervious Cover (%)	Urban runoff concentrations (mg/L)			Annual loading rates * (lb/acre)		
			TOC	TP	TSS	TOC	TP	TSS
Impervious cover (Simple Method)	LDR (<1du/acre)	12	11.9	0.4	100	13	0.4	112
	MDR (1-4 du/acre)	25	11.9	0.4	100	23	0.8	196
	HDR (>4 du/acre)	45	11.9	0.4	100	39	1.3	324
	Commercial	72	11.9	0.2	75	59	1.0	373
	Roadway	80	11.9	0.5	150	65	2.7	822
	Industrial	53	11.9	0.4	120	45	1.5	450
Pervious cover (export coefficients)	Cropland	-	-	-	-	30	2.0	300
	Pasture / Range	-	-	-	-	10	0.7	100
	Forest	-	-	-	-	30	0.2	100

Source: Based on data from Caraco 2001a.

*Loading rates for impervious cover are calculated using Simple Method assuming 35" annual rainfall.

Secondary Sources

The WTM was used to predict pollutant loads from secondary sources including channel erosion, failing septic systems, concentrated animal feeding operations (CAFOs), wastewater treatment plants (WWTPs), CSOs, and SSOs. Calculation methods for secondary sources vary depending on the source. Brief model descriptions, assumptions, and data sources are provided below. More detailed descriptions are provided in Caraco (2001).

Wastewater Treatment Plants

Pollutant loads from wastewater treatment plants are calculated based on median effluent concentrations and median per capita flow rates. Default effluent TOC, TP, and TSS concentrations for a conventional secondary treatment plant are 25, 5, and 30 mg/L, respectively. Default effluent TOC, TP, and TSS concentrations for a tertiary treatment plant providing filtration and biological nutrient removal are 20, 1.5, and 5 mg/L, respectively. Default wastewater flow rate is 165 gal/capita/day. The population served is based on the population density of the watershed and the fraction connected to sewers.

SSOs

The WTM estimates the SSO load as a product of total flow from SSOs and concentrations for raw sewage. WTM assumptions are documented in Caraco (2001). Model default values assume 140 SSOs occur per 1,000 miles of sewer per year, and an average SSO volume of approximately 90,000 gallons.

CSOs

WTM methodology and assumptions for estimation of CSO loads are documented in Caraco (2001). Model default values assume 65 CSO events per year. CSO volume is calculated as the runoff caused by the difference between the median storm event depth and the rainfall depth that causes CSOs (assumed to be 0.1 inch). The runoff volume from this storm event is determined using the Simple Method. The load is the product of this volume, the number of CSO events, and typical concentrations in CSOs.

Failing Septic Systems

Failing septic systems are clearly a significant source of nitrogen, and may contribute to TP and TOC loads as well. However, defining the magnitude of the load is problematic, in part because the definition of failure varies. Failures range from surface discharge of untreated wastewater to subsurface discharge of minimally treated wastewater. The model default assumption is that 10% of all septic systems are failing, and that 10% of these failing septics (1% of total) are complete failures, which result in a direct wastewater discharge. The remainder (9% of total) contributes to subsurface flow only. Subsurface loads of TP, TOC, and TSS are assumed to be fully retained before reaching the receiving water (delivery ratio of zero). The default model assumption is thus that 1% of all septic systems leads to direct wastewater discharge. Based on these default assumptions, failing septics are expected to make a negligible contribution to total pollutant loads in most water supply watersheds.

Channel Erosion

It is well established that increases in impervious cover due to urbanization and development contribute to channel erosion (see e.g. Schueler 2000, Caraco 2000). In the WTM the underlying model for channel erosion is based on an empirical relationship between watershed impervious cover and enlargement ratio, where the latter is defined as the ratio of post- to pre-disturbance channel cross-sectional area. This conception of channel erosion assumes that a stream responds to an increase in impervious cover by increasing its cross-sectional area over a long period, say 50 years. The model distributes the total solids load from channel erosion evenly throughout the response period. Further model and scenario assumptions include:

- The impervious cover: enlargement ratio relationship is based on alluvial soil, and does not apply to channels with rock beds or banks.
- Channel erosion is computed for the fraction of channel length draining urban land cover. Channel length is calculated based on an assumed drainage density of 4 mi/mi^2.[4] Impervious cover fraction is normalized to urban land area, not total watershed area.
- The original channel cross-sectional area is assumed to be 25 square feet.[5]
- Channel enlargement is assumed to occur over 50 years.
- Effects of pre-existing development: The base case scenario assumes that the stream has already fully responded to the existing development level, and that

---

[4] Typical drainage densities range from 1.5 to 6 mi/mi^2.

[5] Based on Eastern U.S. regional curve relating bankfull channel cross-sectional area to drainage area (Dunne and Leopold 1978, as cited in FISRWG 1998). Drainage area is assumed to be equal to the total urban land use area.

there is no pollutant load from channel erosion. Subsequent growth scenarios assume that the channel responds to the increase in impervious cover from the base case. For example, if the base case contains 1% impervious cover and the growth scenario contains 3% impervious cover, channel erosion calculation in the growth scenario assumes that one-third of the potential channel erosion has already occurred.

- TP and TOC loads from channel erosion are calculated assuming an enrichment factor of 2, soil TP content of 0.15%, and soil TOC content of 1.7% (Caraco 2001).

It is important to note that the channel erosion loads predicted in development scenarios represent the long-term loads that occur after the catchment has reached the specified development density.

Concentrated Animal Feeding Operations

The impact of concentrated animal feeding operations is investigated through the addition of pollutant loads based on the number of livestock in the watershed. The magnitude of the CAFO impact is determined by specifying the animal unit (AU) density within the watershed.[6] Representative animal unit density values for modeling scenarios are based on animal census data from Missouri and Iowa, the frequency distributions of which are averaged and summarized in Table 3.2 below (Missouri UWASC 1998, Iowa DNR 1998).

Table 3.2
Animal unit density for livestock-intensive states

Percentile	animal units per 10,000 acres
0%	500
10%	900
20%	1100
25%	1200
30%	1300
40%	1600
50%	1700
60%	2000
70%	2100
75%	2300
80%	2500
90%	3500
100%	6800

[6] For the purposes of CAFO determination, one animal unit is equivalent to one beef cow, 0.7 mature dairy cows, 2.5 hogs over 25 kg, or 100 chickens with unlimited continuous flow watering systems. This is the EPA animal unit, not the USDA animal unit, which is based on animal live weight.

The impact of concentrated animal feeding operations is modeled using three scenarios:

- Base case 0 AU / 10,000 acres
- Low density 1000 AU / 10,000 acres
- High density 2000 AU / 10,000 acres

Simulation of pollutant export from CAFOs is limited to solids, phosphorus, and organic carbon. Pathogens are not included because of the high degree of uncertainty associated with pathogen production (herd infection prevalence and pathogen shedding intensity) and removal/inactivation (during storage, land-application, and export from the land surface).

Pollutant export from CAFOs is calculated as the mass of pollutant generated per animal unit, multiplied by a delivery factor. Pollutant generation is based on typical manure characteristics compiled by ASAE (1999). TOC data for livestock waste is limited. TOC production is calculated from average manure BOD5 characteristics (ASAE 1999), assuming a BOD: TOC ratio of 0.66.[7] Model default values for annual pollutant generation by CAFOs are summarized in Table 3.3 below.

Table 3.3
Model default values for annual pollutant generation by CAFOs

	animals/AU	fixed solids	TP	TOC
		(lb/animal unit/yr)		
Beef	1	380	27	719
Dairy	0.7	715	34	880
Hog	2.5	308	22	588
Chicken*	100	493	33	556

*represents average of broiler and layer

Pollutant export from CAFOs varies widely depending on the region, facility size, type of livestock produced, and waste collection and disposal system. EPA (2001a) provides an excellent national-level, sector-specific review of CAFO waste. Poultry manure is often mixed with litter and is consequently relatively dry, collected seasonally, and stored in stacks prior to disposal. Hog and dairy manure is wet, and confinement areas may rely on flush systems for washing or waste collection, so waste is commonly stored in waste storage ponds prior to disposal. Anaerobic lagoons may also be used to provide treatment prior to disposal.

Though new methods of ultimate disposal (e.g. pelletization, composting) are the subject of increasing research, the vast majority of livestock waste is land applied through spreading or spraying, depending on the solids content of the manure. The environmental impact of this land application depends largely on the manure application rate (e.g. lb P/acre), which depends on the herd size and the availability of suitable crop or pastureland for land disposal. TP export also depends on the timing of land application relative to rainfall, the nutrient uptake characteristics of the crop or pasture grown at the land disposal site, and the runoff and leaching potentials of the soils. McDowell et al. (2001) review factors affecting the release of soil phosphorus into solution and its subsequent transport in surface and sub-surface runoff. Kellogg et al. (2000)

[7] Median value for agricultural waste from data cited in San Diego-McGlone et al. (2000).

provide a national-level assessment of the potential for animal waste in a given county to exceed crop uptake requirements. Roughly 20% of the farm-level excess phosphorus produced in 1997 exceeded the county-level crop uptake assimilation capacity.

CAFO simulations consider only pollutant export due to routine runoff of manure pollutants from land application sites. Pollutant export from CAFOs due to runoff from land application sites is based on a delivery ratio. Gross delivery ratio estimates range from 5% for the Neuse River Basin (NCDEHNR (1993) to a national value of 15% (Kellogg et al. 2000). Model simulations assume a 10% delivery ratio for phosphorus and solids, and 5% for organic carbon to reflect attenuation during storage and land application. These ratios are expected to be representative of CAFOs with inadequate land area for manure application at agronomic rates. CAFOs with sufficient land application area may export less TOC, TSS, and TP to receiving waters. Pollutant export due to catastrophic failure (e.g. storage facility failure, flooding, or intentional discharge) is inherently difficult to predict, and is not considered in model simulations. A detailed nationwide summary of documented CAFO discharges is provided in USEPA (2001d).

Best Management Practices

Urban BMPs

Annual pollutant load reduction efficiencies for structural urban BMPs are based on values reported in Caraco 2001a. Model default values are rounded to the nearest 5% and summarized in Table 3.4.

Table 3.4
Typical pollutant removal efficiencies for structural urban BMPs

BMP	TSS	TP	TOC	Bacteria
Dry pond	5%	20%	10%	10%
Dry extended detention pond	60%[1]	20%	30%[1]	60%[4]
Wet pond	80%	50%	45%	70%
Wetland	75%	50%	20%	80%
Water quality swale[2]	80%	35%	70%	0%[4]
Filters[3]	85%	60%	55%	35%
Infiltration trench	90%[4]	70%	80%[1,4]	90%[4]

1. Performance based on fewer than five data points
2. Refers to open channel practices designed for water quality
3. Excludes vertical sand filters and filter strips
4. Removal rates adjusted based on best professional judgment

Source: Based on data from Caraco 2001a.

Agricultural BMPs

Annual pollutant load reduction efficiencies for agricultural BMPs are based in part on median values from the EPA Chesapeake Bay Program's draft *BMP Effectiveness Database* (Dillaha et al. 2000). Practices supported by few studies are supplemented with additional studies from the literature. Model default values are rounded to the nearest 5% and summarized in Table 3.5. Several of the agricultural BMPs included are multi-component management systems (e.g. integrated pesticide management) whose performance can be expected to vary widely depending on the specific practices implemented.

Table 3.5
Typical pollutant removal efficiencies for agricultural BMPs

BMP	TSS	TP	TOC	Bacteria	Atrazine
Residue management - no till/strip till	80% [1]	70% [1,2]	40% [4]	*	40% [2]
Residue management - reduced till	60%	50%	30% [4]	*	30% [1,2]
Integrated pesticide management [3]	*	*	*	*	25% [2]
Nutrient management system [3]	*	30%	NA	*	*
Grazing management system [3]	80%	60% [2]	40% [4]	70% [2]	*
Animal waste management system [3]	60%	40%	30% [4]	75% [2]	*
Buffer strip	80%	65%	50% [2]	50% [2]	90%
Settling basin	60%	40%	35% [2]	65% [2]	5% [2]
Wetlands	85%	50%	-20% [1,2]	75% [2]	60% [2]

* BMP not applicable for this pollutant
NA Performance data not available
1. Removal rates adjusted based on best professional judgment
2. Performance based on fewer than five data points
3. Actual performance varies widely according to specific practices employed
4. No TOC removal data available. TOC removal assumed to be 50% of TSS removal efficiency

MODELING SCENARIOS

The effects of land use change and BMP implementation on point and nonpoint source pollutant loads were investigated by comparing pollutant export from a base case watershed with pollutant export under an alternative land use or BMP scenario. Results are thus presented as the percent increase or decrease in pollutant load from the base case to the alternative scenario. Three base case watersheds characterized by dominant land use type were created for study: undeveloped (forested), agricultural, and mixed. Land use percentages for these three watershed types were drawn from average land use data for 191 USGS Phase I NAWQA program sites representing primarily undeveloped, agricultural, or mixed land use watersheds. Average land use data for the NAWQA sites is summarized in Table 3.6.

Table 3.6
Average watershed land use distribution

land use type	Number of sites	urban	Agricultural	range	forest	other
		(percent)				
Forested	28	1	4	13	76	6
Agricultural	75	2	73	6	17	3
Mixed	88	4	32	16	43	6

Source: Based on data from USGS 2000.

Land cover for alternative scenarios was calculated based on hypothesized changes in watershed population density. National values for the population density of drinking water supply watersheds were calculated from a USGS data set detailing population and drainage area for 468 drinking water supply intakes located on RF1 reaches (Alexander 2001). Further description of USGS methodology is provided in Alexander (1999). The frequency distribution of watershed population density differs slightly from that of nationwide county population density, as shown below in Table 3.7. Watershed population density values were judged to be more descriptive of water supply source areas than county-level data, and were therefore selected as the basis for default population density values used in watershed scenarios. Modeling scenarios examine five different population density levels, as shown below.

Table 3.7
U.S. population density frequency distributions and model default values

Percentile	Model Scenario name	Population density (pop/km^2)		
		Model default value	By source water area (n=468)*	By county (n=3192)†
0%			0.1	0.0
10%	low density	5	4.5	1.6
20%			7.7	4.4
25%	first quartile	10	12	6.2
30%			16	7.7
40%			22	11
50%	Median density	25	26	15
60%			33	20
70%			43	30
75%	third quartile	50	49	37
80%			56	49
90%	high density	75	74	112
100%			378	20,239

**Source:* Based on data from Alexander 2001.
†*Source:* Based on data from US Census Bureau 2000.

Residential land use cover is calculated from population density based on housing densities of 2.0, 0.5, and 0.1 acres per dwelling unit for low, medium, and high density residential housing, respectively. Dwelling unit population is assumed to be 2.6 people, based on 2000 census values. Commercial, industrial, and roadway land use percentages are assumed to be 1.0%, 0.5%, and 0.5%, respectively, for median population density watersheds (25 pop/km^2). Roadway value is based on EPA estimates of total US road surface area,[8] and commercial/industrial values are based on best professional judgment. Commercial, industrial, and roadway land use percentages for lower or higher population density watersheds are scaled

[8] USEPA (1999a), based on data from FHA (1998) Table HM-60.

linearly from their default values according to the percent change in population density. Crop and pasture land use percentages are set at desired values, and forest cover is calculated as the balance. Population density and land use data for base case watersheds is summarized in Table 3.8 below.

Table 3.8
Population density and land use of model base case watersheds

Watershed type	population density (pop/km^2)	LDR	MDR	HDR	Comm'l	Roadway	Ind'l	Crop	Pasture	Forest
Forested	5	1.6%	0.0%	0.0%	0.2%	0.1%	0.1%	5%	10%	83%
Agricultural	10	1.6%	0.4%	0.0%	0.4%	0.2%	0.2%	75%	10%	12%
Mixed	25	2.6%	0.6%	0.1%	1.0%	0.5%	0.5%	30%	15%	50%

CHAPTER 4
RESULTS: WATERSHED MODELING

INTRODUCTION

Watershed model runs were conducted to investigate: (1) the relative contributions of different primary and secondary sources to total pollutant loads, and (2), the effects of land use change and BMP implementation on pollutant export. The watershed model runs may be broadly divided into development scenarios, urban scenarios, and agricultural scenarios. Development scenarios quantify changes in pollutant loads due to population growth and land development, as well as the potential for BMPs to reduce these loads. Urban scenarios quantify pollutant loads from more highly urbanized watersheds, and the potential for structural urban BMPs to reduce these loads. Agricultural scenarios focus on pollutant loads due to agricultural practices and concentrated animal feeding operations, and assess the potential for BMPs to reduce these loads.

For each watershed model run, salient observations are provided in the text below. Runs are referenced according to a letter and a number (e.g. F4), which corresponds to the scenario details (land uses, population density, BMP application) and results presented in Appendix A. Results are presented as percent change in total annual export of TOC, TP, or TSS, relative to the model run cited in the "base case" column. Runs followed by a decimal place and number (e.g. F4.1) are BMP implementation runs.

DEVELOPMENT SCENARIOS

The 1997 Natural Resources Inventory documents that forest land and cultivated cropland make up the majority of land being developed: since 1982, over 60% of newly developed acreage has been converted from forest or cropland. Two major population growth and land development scenarios are therefore considered below: growth in a relatively pristine forested watershed, and growth in a primarily agricultural watershed. These scenarios are examined by making changes to the base case forested and agricultural watersheds described in Table 3.8.

As described previously, population density is used as the primary variable for defining alternative watershed scenarios that illustrate the effects of population growth and development. Increases in population density are assumed to have three effects: increased pollutant load from WWTPs, conversion of pervious cover to impervious cover, and channel erosion due to increases in impervious cover.

Development effects were examined using the base case forested watershed (see Table 3.8). Forest land is the major land use (83%) in the watershed, and contributes roughly 90% of TOC, 75% of solids, and 50% of phosphorus, as shown in Figure 4.1 below. At a low population density of 5 pop/km^2, impervious cover contributes less than 5% of all pollutant loads. Cropland and pasture are expected to be significant sources of phosphorus, 30% and 20% respectively.

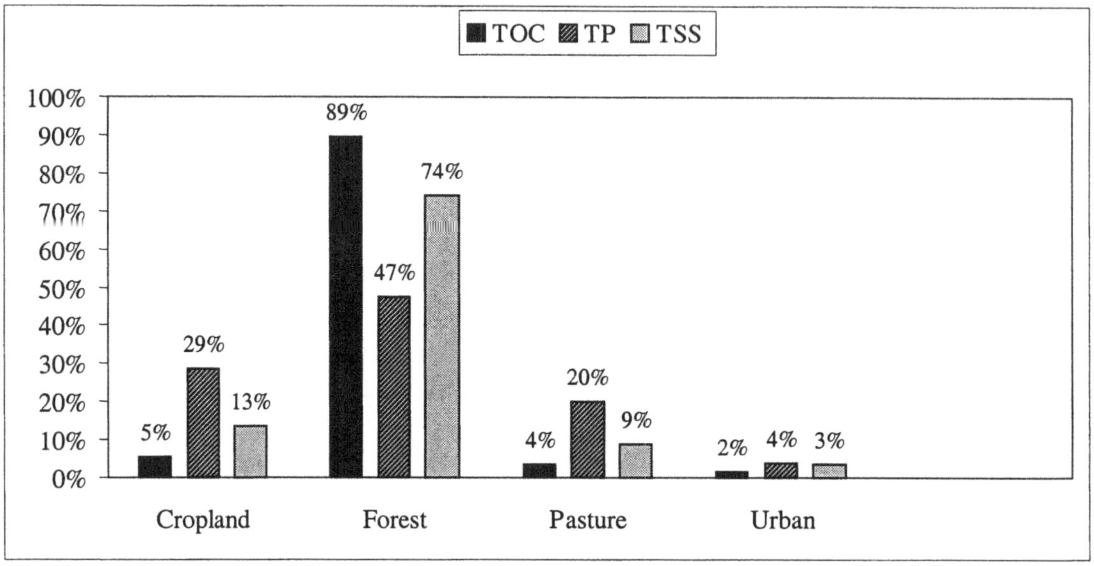

Figure 4.1 Pollutant load contributions in base case forested watershed

Population Growth in a Forested Watershed

The effects of population growth in a relatively pristine forested watershed were investigated by increasing the population density of the base case watershed from 5 pop/km^2 to two different levels representing moderate and high degrees of growth. Within each growth scenario additional runs were conducted to illustrate the effects of housing density, WWTP loads, and channel erosion. Scenario details and load changes are listed in Table A.1.

The moderate growth scenario (run F2) increases the population density to the 25th percentile value of 10 pop/km^2. Negligible (<5%) change in pollutant loads is observed due to changes in land cover alone, regardless of housing density. Default pollutant runoff concentrations are the same for different housing density levels, so the effects of housing density are limited to the amount of forest land converted to impervious cover. For a predominantly forested watershed, moderate population increases have minimal effect on the fraction of forest cover, so little direct effect of land use conversion is observed. Forested land cover remains the dominant source of pollutant loads.

However, significant changes are observed when the effects of a WWTP and channel erosion are included. A WWTP serving the new population is predicted to increase TP loads by 17% (F3). If the effects of channel erosion due to increased impervious cover are included (F4), annual TP and TSS loads are expected to increase by over 50% and 40%, respectively. Combined effects increase TOC loads by 6%. These scenarios indicate that moderate population growth in a forested watershed can lead to significant increases in TP and TSS loads, due primarily to the combined effects of channel erosion and WWTP discharges.

BMPs show moderate ability to reduce these effects. Upgrading facilities at the WWTP to provide tertiary treatment (F4.1) provides a 7% reduction in TP load relative to the development scenario. Tertiary treatment, combined with strategies that reduce by 20% the impervious cover resulting from population growth, reduce TOC, TP, and TSS loads relative to the development scenario by 2%, 14% and 8%, respectively (F4.2).

The high growth scenario demonstrates the effects of increasing the population density of the base case watershed from 5 pop/km^2 to the median value of 25 pop/km^2. It is assumed that the population is evenly divided between low, medium, and high-density residential housing, and that new population growth is served by a conventional WWTP. Model predictions indicate large increases in TP and TSS due to WWTP discharges and channel erosion (F5). Tertiary treatment at the WWTP combined with a 20% reduction in impervious cover (F5.2) provides load reductions relative to the non-BMP case (F5) of 6%, 27%, and 19% for TOC, TP, and TSS, respectively.

Population Growth in an Agricultural Watershed

Two levels of population growth in the base case agricultural watershed are examined, moderate (25 pop/km^2) and high growth (50 pop/km^2), summarized in run series R1 – R5 (Table A.1). Trends are similar to the forested watershed. Population increase leads to increased phosphorus loads from WWTPs, and channel erosion due to increased impervious cover leads to substantial increases in TOC, TP, and TSS. Tertiary treatment and reduced impervious cover are predicted to reduce developed loads of these parameters by 15 – 35% (F3.1, F5.1).

URBAN LOADS AND BMPS IN A DEVELOPED WATERSHED

The base case mixed use watershed (population density 25 pop/km^2) is used to investigate the effects of SSOs, CSOs, and urban structural BMPs. Load reductions due to non-structural urban BMPs such as street sweeping, storm sewer/catch basin cleaning, land reclamation, and illicit connection removal were not quantified, both because the pollutant loads addressed by these practices are difficult to quantify, and because limited monitoring data is available to estimate practice efficiency.

Major land uses in the base case watershed include forest (50%), cropland (30%), pasture (15%), and urban (5%) (see Table 3.8). The relative contribution of these land uses to pollutant loads is shown in Figure 4.2 below. Cropland contributes over half the TP and TSS and one-third the TOC. Forest contributes half the TOC and one-third the TSS. A conventional WWTP serving 100% of the population contributes one quarter of the TP load. Pasture and urban land uses contribute less than 10% of all constituents.

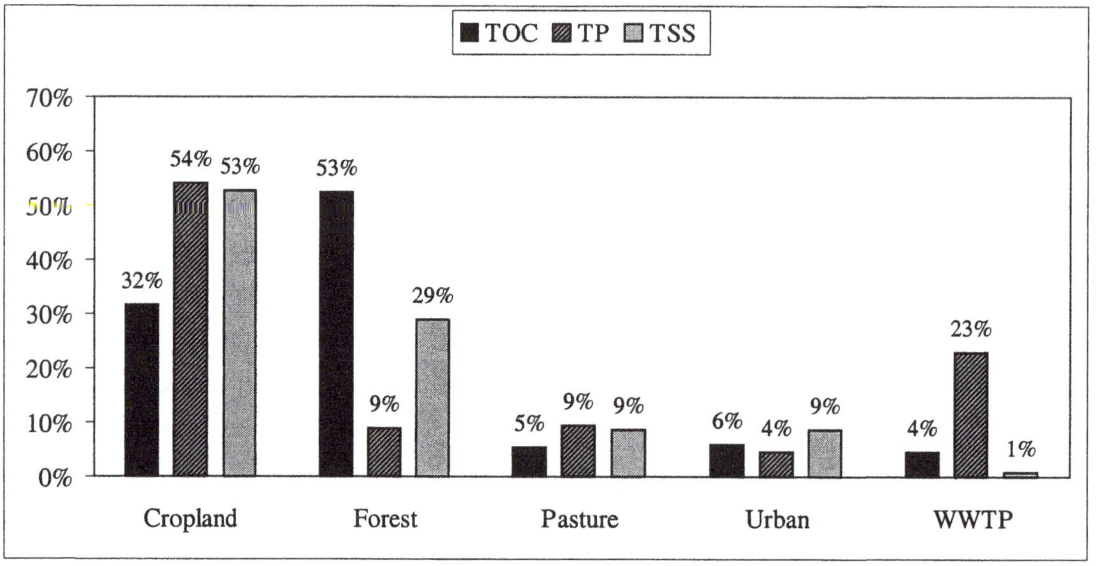

Figure 4.2 Pollutant load contributions in base case mixed land use watershed

SSO loads

In the base case mixed use watershed, using model default SSO assumptions and assuming a sewerage density of 10 miles main/mi^2 of sewershed, TOC and TP loads are predicted to increase by <1%, with no change in TSS. Due to the small volume of total load this represents, load reductions due to SSO BMPs such as blockage removal, pipe lining, or capacity increases are not considered.

CSO loads

In the base case mixed use watershed, using model default CSO assumptions, and assuming the sewershed includes all impervious cover acreage, TOC, TP, and TSS loads are predicted to increase by <2%, with no change in TSS. Due to the small volume of total load this represents, load reductions due to CSO remediation programs are not estimated.

Structural Stormwater Management Practices

Structural stormwater BMPs can achieve relatively high pollutant reduction efficiencies (see e.g. Table 3.4). However, since the urban/impervious areas in which they are applied may contribute a relatively small fraction of the total pollutant load, their ability to provide significant reductions in total watershed export of TOC, TSS, and TP may be limited. In the base case mixed land use watershed, for example, urban/impervious land uses (residential, commercial, industrial, and roadway) contribute less than 10% of total watershed export of TOC, TP, and TSS (see Table A.1).

Because of the small contribution of urban NPS to total pollutant loads, even high efficiency BMPs such as wet ponds provide little reduction in total loads in the watersheds modeled. Wet ponds capturing 100% of the runoff from 50% of the impervious cover in this watershed reduce TOC, TP, and TSS loads by 1%, 1%, and 3%, respectively. The same scenario in a high population density (75 pop/km^2) mixed use watershed reduces TOC, TP, and TSS loads by 4%, 2%, and 9%, respectively. Similar results are observed for other, lower efficiency, urban structural BMPs. In practice, total pollutant load reductions due to urban structural BMPs would be less, due to limitations on the amount of impervious surface area that can be realistically fitted with BMPs, inability to capture 100% of runoff, and inefficiencies due to irregular maintenance. This assessment of structural stormwater BMPs is based strictly on their ability to provide significant reductions in total loads of TOC, TP, and TSS in typical drinking water supply watersheds. It does not account for their potential benefits in more highly urbanized water supply watersheds where urban land use is a more significant contributor to total TOC, TP, and TSS loads, nor does it account for their ability to remove pollutants other than those considered in the current study.

AGRICULTURAL SCENARIOS

Cropland

Agricultural and CAFO loads and load reductions were examined using the base case agricultural watershed. Cropland is the major land use (75%) in the watershed, and contributes 80% of TOC and roughly 90% of TP and TSS loads, as shown in Figure 4.3 below.

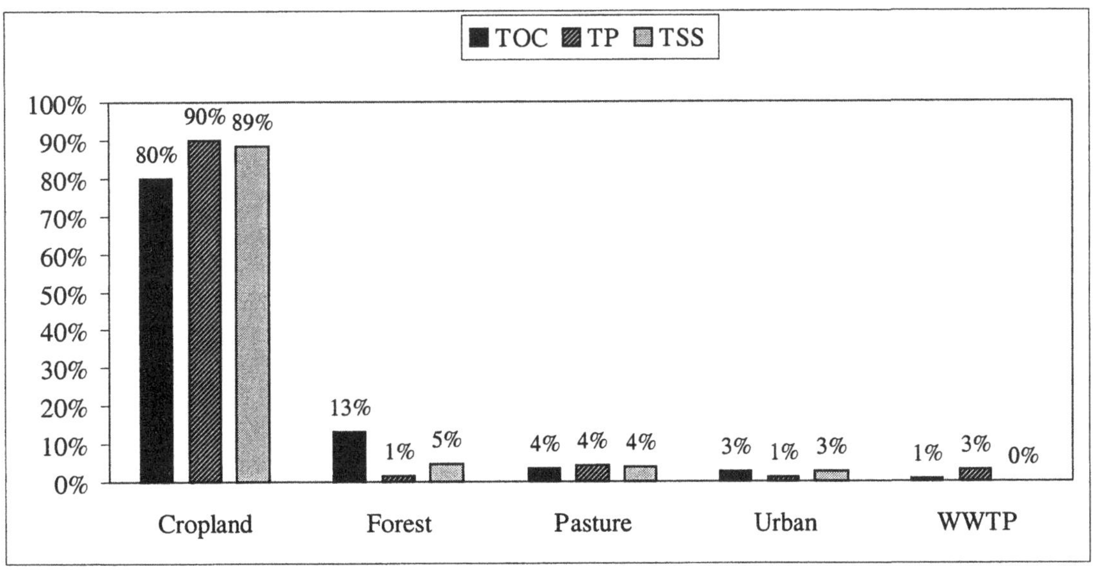

Figure 4.3 Pollutant load contributions in base case agricultural watershed

Load reduction potential of selected agricultural BMPs was investigated at varying levels of implementation. BMPs and pollutant removal efficiencies are summarized in Table 3.5, and run details for cropland runs A1 – A45 are provided in Table A.2. Since the model default removal efficiencies for most of the agricultural BMPs are roughly the same (e.g. 60-80% for solids, 40-60% for TP, and 30-40% for TOC), their load reduction potential depends primarily on the fraction of cropland they treat.

Effects of BMP implementation were investigated at 10%, 25%, and 50% of cropland acreage. At 10% implementation agricultural BMPs such as buffers and conservation tillage provide roughly 5% reduction in TOC, TP, and TSS loads. At 25% implementation level, buffers provide 20%, 29%, and 35% reduction in TOC, TP, and TSS loads, respectively. Settling basins and wetlands are investigated at low implementation levels, since in practice they are not used to treat large portions of the agricultural landscape. Improved nutrient management is predicted to achieve modest reductions in TP export (<15%) even at high levels of implementation (50%).

Combinations of BMPs can provide substantial reductions in pollutant loads. Conservation tillage (no-till) combined with buffer strips provides a roughly 15% reduction in TOC load and 20% reduction in TP and TSS loads.

Buffer strips and conservation tillage were selected for cost-benefit analysis because they are two widely promoted, funded, and implemented agricultural BMPs. Furthermore, they have the greatest potential to treat large portions of the agricultural landscape. BMPs such as wetlands and settling basins are site-specific practices that can be effective in reducing loads from priority pollutant source areas. However, these benefits are not captured in a screening-level export coefficient model. Nutrient management is not considered in the cost-benefit analysis because the impact of reduced phosphorus loads is not quantified in the WTP model.

Buffer strips and conservation tillage were investigated at implementation rates corresponding to load reduction scenarios in the WTP model. Run A5 considers 35% implementation of buffer strips, yielding roughly 25% and 15% reductions in TSS and TOC loads, respectively. Run A7 considers 15% implementation of no-till cultivation, yielding 10% and 5% reductions in TSS and TOC loads. Run A10 considers roughly 55% implementation of residue management, yielding roughly 40% and 20% reductions in TSS and TP loads. Further discussion of the economic benefits of these implementation scenarios is provided in Chapter 7.

Pasture

Impacts of grazing animals and grazing management BMPs is considered in a variant of the agricultural base case watershed. This watershed, 35% pasture and 50% cropland, is predicted to generate 15 – 20% reductions in TOC, TP, and TSS loads relative to the base case agricultural watershed due to lower pollutant export coefficients. Improved grazing management at high levels of implementation (50%) yields less than 10% reductions in pollutant loads relative to the base case pasture watershed.

CAFOs

CAFO impacts are investigated in Runs A30 – A45 (Table A.3). Regardless of animal type (beef, dairy, swine, or poultry), one CAFO in the 10,000-acre hypothetical watershed (equivalent to a roughly 15th percentile animal unit density level) is predicted to generate significant increases in TOC (10-15%) and TP (15-20%), while loads from two CAFOs (60th percentile) are proportionally greater (see runs A30 to A37). Solids contributions are negligible. Animal waste management systems and wetlands are predicted to provide 5% to 20% reductions in TOC and TP loads (Runs A38-A45).

CHAPTER 5
METHODS: WATER TREATMENT PLANT MODELING

INTRODUCTION

This chapter presents a brief review of water treatment plant simulation models, with particular focus on the OTTER model (WRc/D&B 2001). It also describes the calibration, validation, and use of OTTER to simulate the effects of changes in solids and organics loads on four WTPs.

LITERATURE REVIEW

Introduction

While process simulation software is common in the wastewater treatment industry, the development and use of simulation packages for potable water treatment is relatively new. Use of computational fluid dynamics for hydraulics design and optimization is growing, but integrated computer simulation of individual unit processes for water quality optimization is still uncommon. Though the development of analytical and empirical expressions for individual unit processes dates to the birth of the industry and continues today, there is to date only one commercially available dynamic simulation program that integrates individual process models to allow for simulation of entire water treatment plants.

A description of the theoretical foundations and development of analytic expressions for individual water treatment processes is beyond the scope of this review. The interested reader is referred to AWWA's *Water Quality and Treatment* (1999) for detailed process descriptions. This review describes instead the capabilities and limitations of currently available water treatment plant simulation software, and provides a brief description of the underlying unit process models upon which they rely.

Steady State Programs

EPA Water Treatment Plant Model

The EPA Water Treatment Plant Model (USEPA 1994b) is a steady state model developed to predict disinfection byproduct (DBP) formation, natural organic matter (NOM) removal, and disinfectant decay. It was developed in 1992 to support the Disinfectant/Disinfection Byproducts (D/DBP) negotiation process and predict the effects of process modifications on DBP formation. The model algorithms are regression equations developed from water quality databases (see Harrington, Chowdhury, and Owen 1992, and Amy, Chadik, and Chowdhury 1987).

The model is currently under revision (Summers et al. 1999) in order to reflect the development of new algorithms for inactivation, DBP formation, disinfectant decay, and additional unit processes. The revised model (Version 2.0) is expected to provide for the modeling of multiple chlorination points, parallel treatment trains, and the distribution system, and will include improved algorithms for:

- Coagulation
- Softening
- GAC
- Membranes
- Giardia inactivation
- THM and HAA formation - total and individual species
- Chlorine and chloramine decay

The revised model will also include a semi-empirical sorption model (see Edwards 1997) for TOC removal during coagulation. Tseng and Edwards (1999) extend the capabilities of this model to predict pH in open and closed systems and to account for the mixing of two raw water sources.

WatPro

WatPro (Enviromega 2000) provides steady state simulation of DBP formation, giardia/virus inactivation, organic matter removal, disinfectant decay, and pH. WatPro relies largely on EPA algorithms, provides a graphical user interface, and is relatively inexpensive.

Dynamic Programs: OTTER

OTTER (WRc/D&B 2001) is a dynamic water treatment plant simulation program that consists of mechanistic and empirical models for most water treatment unit processes linked together to allow simulation of treatment plant operations and finished water quality over time. The program allows the user to build and calibrate a model treatment plant based on actual plant data, then use the calibrated model to investigate different scenarios such as raw water quality changes or unit process modifications. One of the principal strengths of this modeling package is that individual unit process algorithms are linked together to account for the effects of upstream processes on downstream processes. Coagulant dose, for example, influences floc size, which in turn influences clarification performance and filter performance. The unit process models are described in detail in the OTTER user documentation. Brief descriptions of the processes simulated for this project are provided below.

pH adjustment

The pH model (Ainsworth and Jones, 1979, and Ainsworth, 1979) is based on the fundamentals of the carbonate system in water, and therefore does not require calibration. It can be used either to calculate the pH, hardness, and alkalinity resulting from a given chemical dose, or it can calculate the chemical dose required to achieve a target pH. The model also calculates the equilibrium concentration of dissolved metals (aluminum, iron and manganese), as well as the potential for calcium carbonate scale deposition, dezincification, and plumbosolvency. The model assumes that there is no precipitation of calcium carbonate and that the system is closed to the atmosphere.

Data requirements for the pH model include influent pH, temperature, hardness, alkalinity, and conductivity. If the model is used to predict pH, chemical dosage data is required; if the model is used to predict chemical dosage, target pH is required.

Coagulation

The OTTER coagulation unit predicts: color and DOC removal as a function of coagulant dose and pH, the effect of coagulant dose on pH, coagulant residuals, and suspended solids concentration of process effluent. The organics removal model (Edwards 1997) divides TOC into particulate, adsorbable soluble, and non-sorbable soluble fractions. Removal of the adsorbable fraction is based on a Langmuir isotherm describing the equilibrium between DOC in solution and DOC adsorbed to the surface of metal hydroxide floc. The non-adsorbable fraction is calculated as a linear function of raw SUVA. The final organics removal model contains six calibration parameters, suggested values for which are provided by Edwards (1997).

The effect of coagulant dose on pH and residuals production is based on underlying aquatic chemistry, and varies according to coagulant type. Coagulants covered in the model include alum, PACl, ferric sulfate, ferric chloride, polymer ferric, and ferral. The effect of coagulant type on organics removal is accounted for in the calibration process. Jar test data can be used to guide model calibration. The effect of polymer addition is also accounted for in the calibration of the organics model. Applied polymer dose is also added to the suspended solids stream.

Coagulant dose can be supplied based on plant data, or can be predicted. Otter predicts coagulant dose as a simple, linear, additive function of influent pH, color, turbidity, temperature, alkalinity, and conductivity. Coefficients are calculated by the user based on regressions of historical coagulant dose and water quality data. Alternatively, the user may develop dose algorithms outside the model, and use predicted dosage as model input (see below for further detail on this approach).

Limitations and assumptions of the Otter coagulation unit include:

- Does not account for the effects of temperature on organics removal
- Assumes that the coagulant is dispersed optimally in the water stream
- Does not describe the effect of coagulant dose and mixing conditions on the filterability or settleability of particles (though it is possible to manually specify the fraction of non-settleable and non-filterable solids)
- Calibration is data-intensive due to the large number of calibration parameters

Flocculation

The OTTER flocculation model predicts a uniform effluent floc size from each floc basin as a function of coagulant dose, influent floc size, and maximum floc size. The maximum floc size is calculated as a function of mixing energy (G), as described by Parker et al. (1972). Two calibration parameters are used for maximum floc size prediction, and two for effluent floc size prediction. Effluent floc size affects the performance of downstream clarification processes. The model does not account for the effects of residence time or mixing energy on floc growth, and describes only monodisperse suspensions.

More rigorous attempts to model flocculation are documented in AWWA (1999). Dharmappa et al. (1994), for example, model flocculation using Friedlander's 1977 formulation of Smoluchowski's analytical expression (1917). Thomas, Judd, and Fawcett (1999) describe the six constraints proposed by Smoluchowski's assumptions, and review attempts to address each of these limitations. The assumptions of the Smoluchowski model are:

- All particle collisions lead to attachment
- Fluid motion is limited to laminar shear
- Monodisperse suspension
- No floc breakup occurs
- Particles are spherical in shape and remain so after collision
- Collisions take place only between two particles

Conventional and plate sedimentation

The OTTER horizontal flow clarifier model predicts the effluent solids concentration resulting from discrete (Type I) settling using the Stokes settling velocity. The model accounts for the dependence of water density and viscosity on temperature, and describes particle density as a function of floc size using the model proposed by Lagvankar and Gemmell (1968). The model does not account for flocculent (Type II) settling. The model assumes that all particles are equally well removed, and that settled solids are continuously removed from the tank. Basin hydraulic performance is described by specifying the equivalent number of CSTRs in series (calculated from tracer data). Dharmappa et al. (1994) account for Type II settling using the description developed by Lawler, O'Melia, and Tobiason (1980).

Dissolved air flotation

The OTTER DAF module predicts the removal of total suspended solids based on tank dimensions and flow regime (equivalent number of CSTRs in series), air dose and bubble size, and the characteristics of the floc particles entering the process. It assumes that no flocculation occurs within the flotation tank, that sludge is continuously removed from the tank and does not affect process performance, and that all particles are equally well removed.

Solids removal efficiency is based in part on the floc flotation constant, which is calculated using the relationship developed by Malley and Edzwald (1991). The model is calibrated by adjustments to floc particle and bubble sizes, as well as the floc attachment efficiency. Floc particle size is adjusted in the flocculation process.

Rapid gravity filtration

OTTER provides two well-documented rapid gravity filtration models, a finite difference model and a logistic model. Both are macroscopic models that ignore the specific mechanisms of particle transport, attachment, and detachment (see e.g. Dharmappa et al. (1994) for a microscopic model), and instead combine a solids mass balance with empirical rate expressions (AWWA 1999). The finite difference model, which uses Adin and Rebhun's 1977 rate expression, is slower to solve and includes the effects of backwashing on solids removal. The logistic model is quicker to run and is from Saatci and Oulman (1980). Both models assume that all components of the total solids concentration in the influent are equally well removed by filtration, regardless of pre-treatment methods. The clean bed headloss is calculated within the model using the Carman-Kozeny equation. Model calibration is required to estimate attachment coefficient, filter capacity, non-filterable solids fraction, hydraulic conductivity, and ripening period. The OTTER backwashing model is based on Amirtharajah (1985).

Model requirements include physical characteristics of the filter and filter media, as well as operational parameters. The model may be set to trigger the backwash of an individual filter at user-specified values of head loss, turbidity, or run time.

GAC adsorption

The GAC model uses a contaminant mass balance to predict the removal of pesticides and organics from the influent water. Model details are found in Hart and Carlile (1992). Equilibrium between adsorbed and non-adsorbed parameters is described with a Freundlich isotherm. Model calibration is based on the Freundlich capacity constant, exponential constant, and rate constant. The model does not account for the competitive effects of NOM or other parameters on the adsorption of a given parameter. The model assumes that regeneration restores the GAC to its virgin state.

Chemical Disinfection

OTTER predicts disinfectant demand, disinfectant decay, and DBP formation for chlorine, chloramines, and chlorine dioxide. The model may be used with existing dose data, or may be used to predict the dose necessary to achieved a desired residual. Changes in pH due to disinfectant addition are also included. Numerous different disinfectant decay models are available, depending on the disinfectant chosen. Five different DBP formation models are available, including the EPA model. Calibration parameters for disinfectant decay and DBP formation depend on the specific model chosen. Disinfection is based on the Hom model (Hom 1972) modified to describe an equal number of CSTRs in series. The CT product is calculated through all chemical disinfection processes.

Ozonation

The OTTER ozone contactor model (Shepherd and Hart, 1993) predicts the impact of water quality on ozone demand and ozone transfer, the effect of ozone dose on organics and pesticide removal, and the effect of the ozone residual on disinfection efficiency. The model also calculates the CT product through the ozonation units. The algorithm is a mass balance model that first calculates the transfer of ozone from the gas phase to the liquid phase, and then calculates the decay of dissolved ozone. The flow regime is assumed to be plug flow.

MODELING METHODOLOGY: OTTER

OTTER was used to predict the effect of changes in raw water quality on the operations, costs, and effluent quality of four WTPs. WTP modeling documentation in this report includes descriptions of the WTPs modeled; model calibration and validation procedures and representative validation results; coagulant dose prediction procedures; and assumptions for WTP simulation runs.

WTP Descriptions

Four treatment plants were selected for modeling to represent a range of processes and raw water quality characteristics. Table 5.1 summarizes the average and typical range of values for major raw water characteristics.

Table 5.1
Treatment plants modeled and source water characteristics

Treatment plant	raw water source	PH	turbidity (NTU)	true color (Hazen)	alk (mg/L)
Conventional: FCWA Corbalis WTP	Potomac River	7.9 (7.0 – 9.0)	15 (max >100)	29 (10 - 80)	84 (30 - 130)
Conv'l w/ GAC filter media: IAWCo Muncie WTP	White River	8.0 (7.5 – 8.5)	10 (max >300)	12 (2 - 63)	246 (90 - 350)
Direct filtration: OCWA Marcellus WTP	Otisco Lake	7.7 (7.0 – 8.0)	2 (max <10)	7 (6 - 8)	117 (100 - 125)
Dissolved air flotation: New Castle NY Millwood WTP	Catskill Aqueduct	6.5 (5.5 – 8.0)	2.5 (max <10)	27 (5 - 100)	11 (5 - 30)

Calibration and Validation Procedures

Unit process size and operational data were collected from each WTP and used to construct an OTTER simulation model. Unit processes simulated in the models include pH adjustment, coagulation, flocculation, sedimentation, dissolved air flotation, granular media filtration, GAC adsorption, chlorination, and ozonation. One year or more of raw, process, and finished water quality data and chemical dosage data was collected for each plant. Annual average water quality and chemical dosage data was used in steady state simulations for rough calibration of individual unit processes to match average process and finished water data. Final calibration was based on short (several weeks) dynamic simulations for selected periods of interest during the year. Model validation was conducted by performing long (one year or more) dynamic simulations and comparing model results with actual plant data. Representative model validation results are presented in Appendix B, Figures B.1 – B.11.

Coagulant Dose Prediction

Water treatment plant models were calibrated and validated using actual plant coagulant dose data. Use of WTP models to investigate the effects of different raw water quality, however, requires that the model coagulant dose be increased or decreased to compensate for changes in the solids and organics concentrations in the modified raw water quality scenario. To this end, coagulant dose prediction algorithms based on multivariable linear regressions were developed for each plant. While more sophisticated techniques such as artificial neural networks (see e.g.

Yu et al. 2000) have been shown to have significantly greater predictive power than linear regressions, these methods are beyond the scope of this project.

Algorithms developed for this project predict coagulant dose as a function of turbidity and an organics parameter (apparent color or true color, depending on plant data). While in many cases better fits to plant dosage data can be obtained by including additional predictor variables (e.g. pH, alkalinity, temperature), higher order variables, and interaction terms, these additional variables may not always be significant.[9] Additional variables also tend to obscure and weaken the frequently observed dependence of coagulant dose on organics and solids loads. Dose prediction algorithms for each WTP are provided in Table 5.2 below.

Table 5.2
Coagulant dose algorithms used for model simulations

WTP	Coagulant dose algorithm	R^2
Corbalis	mg/L Al^{3+} = 1.442 + 0.020*turbidity + 0.019*true color	0.40
Millwood	mg/L Al^{3+} = 2.188 + 0.056*turbidity + 0.004*apparent color	0.06
Muncie	mg/L Al^{3+} = 0.026 + 0.522*ln turbidity + 1.397*ln true color	0.45
Otisco	mg/L Al^{3+} = 1.780 + 0.019*turbidity + 0.052*apparent color - 0.014*alkalinity	0.32

The algorithms developed for these plants account for less than half of the variation in actual plant dosage. The predictor algorithm for the Millwood WTP model in particular accounts for very little of the variation in plant dosage. Millwood alum dosage varies widely in order to maintain a zero streaming current detector reading, and available plant data is insufficient to capture this variation in raw water quality. In general, at least two factors complicate the development of coagulant dosage algorithms based on plant dosage data and raw water quality data. First, critical organics parameters such as DOC, TOC, and UV absorbance are typically monitored infrequently. It is expected that more frequent monitoring for these parameters would improve coagulant dose prediction substantially. Second, it is likely that the plants do not deliver the optimum coagulant dose 100% of the time, which introduces additional error into the dose prediction algorithm.

Dose prediction algorithms were validated by comparing actual plant finished water quality under actual coagulant dosage with model finished water quality under predicted coagulant dosage. Despite the difficulties of coagulant dose prediction with limited organics data, no substantive difference in solids or organics removal performance was observed between finished water quality under actual and predicted coagulant dose conditions.

Assumptions for WTP Simulation Runs

In order to provide a uniform evaluation of the effects of raw water quality changes on WTP performance, several changes were made to the four validated WTP models for use in simulation runs:

[9] Alkalinity was a significant predictor variable in the Otisco algorithm; see Tseng et al. (2000) for a discussion of the impact of alkalinity on coagulant demand.

- Coagulation pH was maintained at the plant average coagulation pH by automatically dosing caustic soda at the point of coagulant addition.
- Finished pH was adjusted by automatically dosing caustic soda to the plant effluent stream to maintain the plant average finished pH.
- Pre-chlorine dosage was automatically adjusted to maintain the plant average pre-chlorine residual at the plant measurement point. Ozone dosage in the Millwood model was automatically adjusted to maintain a residual of 0.8 mg/L leaving the contactors.
- Post-chlorine dosage was automatically adjusted to maintain a 2 mg/L residual leaving the plant.
- Filter backwash triggers were set at 0.2 NTU and the actual plant head loss trigger value.
- An additional 48 hours detention time after plant effluent was added to simulate DBP formation in the distribution system.
- Coagulant dosage in all simulation runs including the baseline run is based on the dosage algorithm for each plant.

CHAPTER 6
RESULTS: WATER TREATMENT PLANT MODELING

INTRODUCTION

Calibrated and validated OTTER models of four treatment plants were used to predict the effects of changes in raw water solids and organic carbon loads on treatment costs, plant operations, and finished water quality. Cost parameters considered include chemical usage, residuals disposal, and power consumption of washwater pumpage. Simulation runs were conducted by scaling the daily values of solids and organics parameters in the baseline raw water quality profiles for each WTP up or down according to the relative change in annual TSS and TOC export in selected watershed scenarios.

Fewer WTP model simulations were conducted than watershed simulations for several reasons. First, similar load changes in many of the watershed model runs allows the use of one WTP run to capture the effects of multiple watershed scenarios. Second, small differences in loads between various watershed scenarios have little practical effect on WTP modeling results. Third, OTTER run time places a practical limit on the number of runs that can be conducted.

Seven sets of alternative raw water quality scenarios were examined at each WTP to assess the plant's response to increases or decreases in organics and solids loads. These scenarios reflect both improvements and deteriorations in raw water quality relative to the base case scenario, as summarized in Table 6.1 below. All OTTER model scenarios span one year and include typical seasonal and storm event water quality fluctuations.

Table 6.1
Raw water quality changes for WTP simulation runs

WTP run	Change	solids	organics
		(percent change from base case)	
1	Base case	-	-
2	Deterioration	25%	25%
3	Deterioration	50%	25%
4	Deterioration	100%	50%
5	Improvement	-10%	-5%
6	Improvement	-25%	-15%
7	Improvement	-40%	-20%
8	Improvement	-50%	-33%

MODELING RESULTS

Results of WTP simulation runs are summarized in Appendix B, Table B.1. Data reported include annual values of:

- Average filter applied turbidity
- Average plant effluent turbidity

- Average filter run time (hours)
- Percent change in total volume of water used for backwashing
- Percent change in average plant effluent TTHM and HAA5 concentrations
- Percent change in pre-oxidant consumption (chlorine or ozone)
- Percent change in coagulant consumption (ferric chloride or PACl)
- Percent change in total mass of sludge produced
- Percent change in post chlorine consumption
- Percent change in pre- and post- caustic soda consumption
- Percent change in treatment costs

Data reported as "percent change" are calculated relative to the base case scenario. Treatment costs quantified include costs of backwash water pumping, solids disposal, and chemical consumption. Treatment costs are calculated based on unit costs selected to represent typical industry values. For a given chemical, the same unit costs are used for each plant in order to allow comparison between plants. Further detail on costs is provided in Chapter 7.

The data presented in Table B.1 indicate that the WTP model predicts realistic changes resulting from raw water quality changes. For each plant, effluent turbidity remains essentially constant across runs, effectively simulating the backwashing of filters as frequently as necessary to maintain effluent turbidity under the plant head loss constraints. Filter run times decrease (backwash frequency and washwater volume increase) under increased solids loads. This effect is minimal for the conventional WTPs (Muncie and Corbalis), reflecting solids removal through clarification units. The effect is more pronounced at Millwood and Otisco where steady, low turbidity sources led to the selection of DAF and direct filtration respectively, processes that are more sensitive to changes in solids loads.[10] Coagulant demand, chlorine demand, and DBP formation are predicted to increase because of increased organics loads. Caustic soda consumption increases with organics and solids to maintain plant coagulation pH under increased coagulant dose conditions.

Some deviations from expected trends are observed. DBP levels are in some cases predicted to increase under decreased organics loading conditions (e.g. Muncie, runs 5 – 7). This is likely the result of the predicted coagulant dose under the alternative scenario providing slightly less efficient organics removal than the baseline scenario.

Model runs indicate that moderate deteriorations in raw water quality (e.g. 25% increase in solids and 25% increase in organics) can lead to treatment cost increases of roughly 10%, while more marked deterioration (100% increase in solids, 50% increase in organics) can increase treatment costs by roughly 15%. Treatment costs are not directly proportional to changes in solids and organics loads.

Model runs also indicate that moderate improvements in raw water quality (e.g. 25% decrease in solids and 15% decrease in organics) can reduce treatment costs by roughly 5%, while more significant improvements (e.g. 40-50% decrease in solids, 25-30% decrease in organics) can reduce treatment costs by roughly 10%. These simulations suggest that routine operating costs do not exhibit high sensitivity to changes in TOC and TSS concentrations (that is, the percent change in operating costs is substantially less than the percent change in pollutant

---

[10] In practice the DAF recycle rate would be increased to maintain floated water turbidity. This scenario was not considered in model simulations but would be expected to increase treatment costs.

loads). The observed changes in operating costs are nonetheless significant from an economic perspective.

Atrazine

The Muncie WTP was selected for modeling primarily to investigate the potential for BMPs to reduce finished water atrazine levels and reduce the costs associated with atrazine removal. The Muncie plant experiences baseline raw water atrazine concentrations in the 0.05 – 0.25 ppb range during winter months, but experiences peaks as high as 10 ppb during rainfall events following pesticide application. The plant uses two treatment techniques to comply with the 3 ppb atrazine MCL. First, PAC is fed starting during the first rainfall event in April and is applied continuously through mid-September. Feed rates are 2 ppm when raw turbidity is under 20 NTU and 5 ppm when turbidity exceeds 20 NTU. Second, the plant uses 24" of GAC in its dual media filters. Plant data indicates that PAC application reduces atrazine concentrations by roughly 30%, while GAC provides an additional 40% reduction. GAC media is replaced every two years.

The Otter model was used to simulate atrazine removal through GAC filters. The Otter software does not currently simulate pollutant removal with PAC, so influent atrazine concentrations were reduced by 30% during PAC application periods. The Otter GAC model predicts removal of pesticides and other organic constituents according to Freundlich isotherms calibrated for each constituent. The model does not currently account for NOM competition for adsorption sites. Atrazine removal at the Muncie plant was validated against plant performance for the period 1/1/99 – 10/1/01. Validation results presented in Figure B.10 indicate a strong correspondence between actual and predicted effluent atrazine concentrations.

Due to limitations in the ability of the watershed modeling tools to provide meaningful predictions regarding the effect of BMP implementation on atrazine export, especially during runoff events, OTTER simulations were not conducted to quantify the effects of reduced atrazine loads on finished water atrazine concentrations.

CHAPTER 7
COST-BENEFIT ANALYSIS

INTRODUCTION

While source water protection measures may reduce pollutant loads and improve raw water quality, it is useful to assess the cost of this water quality improvement relative to the benefits realized at the water treatment plant. Key factors affecting the cost of source protection measures for water utilities include the catchment size, the pollutants of concern, pollutant source location in the watershed, the ability of the source protection program to target sites with high pollutant yields, and the availability of multiple partners to share in the cost of BMP implementation. Benefits of water quality improvement at the water treatment plant are influenced by the size and unit processes of the treatment plant and routine operating costs. Operating costs considered include power consumption, chemical consumption, and residuals handling and disposal.

This chapter describes the methodology and assumptions that were used to quantify (1) the impact of raw water quality changes on treatment plant operating costs; (2) the cost-effectiveness of BMPs as a means of reducing treatment costs; and (3) the fraction of plants that may benefit from promoting BMP implementation. The first two sections describe BMP cost assumptions and WTP cost assumptions, respectively. The third section describes the cost-benefit methodology, and the fourth summarizes results.

BMP COSTS

Costs for some urban and agricultural BMPs are provided in Caraco (2001), as well as USEPA (2000a). Further detail on costs for BMPs selected for cost-benefit analysis is provided below.

Riparian Buffers

Lynch and Tjaden (2000) and Palone and Todd (1998) outline implementation costs of grass and forested riparian buffers in the Chesapeake Bay Region, though much of their discussion is applicable nationwide. Opportunity costs of the land taken out of production comprise a large part of total buffer implementation cost, though these costs fluctuate depending on yields and commodity prices in a given year. Buffer installation costs are another major factor, though these vary widely depending on buffer width, species composition, (e.g. grass vs. forested) and maintenance requirements. Finally, federal and state subsidies under various programs (e.g. CRP, CREP) vary from state to state but may cover a large fraction of buffer costs, depending on specific program details (e.g. incentive payments, installation cost-share, annual rental payments). Because of the high degree of variability in opportunity costs, installation and maintenance costs, and subsidies, costs of riparian buffers are best calculated on a state or site-specific basis.

To capture some of this variability, cost-benefit analysis examines implementation of a grass and legume buffer strip at two different price levels: $140/acre and $50/acre. The higher cost is based on $120/acre/year crop rental payment for corn-soybean rotation, plus installation and maintenance costs, and assumes a 10 year return period and 10% internal rate of return (Nakao et al. 1999). The lower level reflects lower crop rent due to less productive land or lower cost crops. Cost-benefit analysis for riparian buffers also assumes that 50' buffers are implemented on both sides of the stream, and that the watershed has a uniform drainage density of 4 mi/mi^2.

Conservation Tillage

Annual implementation costs for conservation tillage range from $10 to $50/acre/year (Caraco 2001), depending on region, equipment, and cropping system. Cost-benefit analysis assumes a cost of $25/acre/year.

WTP COSTS

WTP operating costs quantified in model runs include chemical consumption, residuals disposal, and energy usage due to washwater pumping. Costs were supplied by the American Water Works Service Company based on their extensive experience.

Chemical Costs

• PACl	$0.10/wet lb (bulk, liquid)
• Ferric chloride (38%)	$0.04/wet lb (bulk, liquid)
• Alum (50% as hydrate)	$0.08/dry lb (bulk, liquid)
• Caustic Soda (50%)	$0.15/dry lb (bulk, liquid)
• Acid	$0.04/lb (bulk, liquid)
• Ammonia (bulk aqua (19%))	$0.21/dry lb
• Chlorine (gas)	$0.14/lb (one ton cylinders)
• Ozone (LOX operating costs)	$15/mgd

Solids Handling

Since residual solids processing and disposal options at a particular plant may include sewer discharge, lagooning, drying beds, land application, mechanical dewatering, monofill disposal, or any combination of these techniques, treatment and disposal costs vary widely. Combined treatment and disposal costs are site-specific and can vary from $100 to over $1500 per dry ton. For modeling purposes it is assumed that the combined cost of treatment and disposal is $100 per dry ton.

Backwash Water

It is assumed that backwash water has no reclaimed value due to recycle, and that pumping costs $0.05/1000 gallons.

METHODOLOGY

A key factor influencing the cost-effectiveness of source protection measures is the relationship between the size of a treatment plant and the size of its watershed. Small WTPs in large catchments may lack the resources to affect watershed changes that will have an impact on treatment costs. Conversely, large WTPs in small watersheds may find it cost-effective to implement BMPs to reduce treatment costs. National data on the size of water treatment plants and the size of their watersheds was used to develop a frequency distribution of the ratio of watershed size to water treatment plant average production. This frequency distribution was used to estimate the percentage of water treatment plants that would benefit from source protection measures as follows:

1. Quantify the annual costs of BMP implementation to achieve a specified reduction in watershed export of organic carbon and sediment. Normalize these costs to catchment size, yielding units of $\$/mi^2/yr$.
2. Quantify the annual benefits of the specified organic carbon and solids load reductions at the treatment plant. Normalize these costs to average WTP production, yielding units of $/mgd/yr
3. Calculate the benefit-cost ratio to determine the break-even ratio of treatment plant size to watershed size (mi^2/mgd). Compare this break-even ratio to the national frequency distribution of the ratio of watershed size to treatment plant size in order to estimate the percent of WTPs that would find BMP implementation would be a cost-effective means of reducing treatment costs.

Relationship between Watershed Size and WTP Size

The relationship between watershed size and WTP size for U.S. utilities was quantified using USGS data for 960 drinking water supplies withdrawing via 1375 intakes (Alexander et al. 1999). The USGS data set summarizes WTP locations and intake characteristics from the USEPA SDWIS database, and water body and land use data from the USEPA RF1 database. Total watershed size was estimated by summing the drainage area above each intake for all intakes feeding a given water treatment plant. Average WTP production was estimated by multiplying the population served by a representative mean consumption rate of 154 gpcd (AWWA, 1996). The ratio of these two numbers, total watershed size to average WTP production, has units of mi^2/mgd. The frequency distribution of this ratio is shown in Figure 7.1 below. This ratio spans a very wide range, roughly seven orders of magnitude. The median value is 170 mi^2/mgd, and 50% of the supplies are between roughly 40 and 1000 mi^2/mgd.

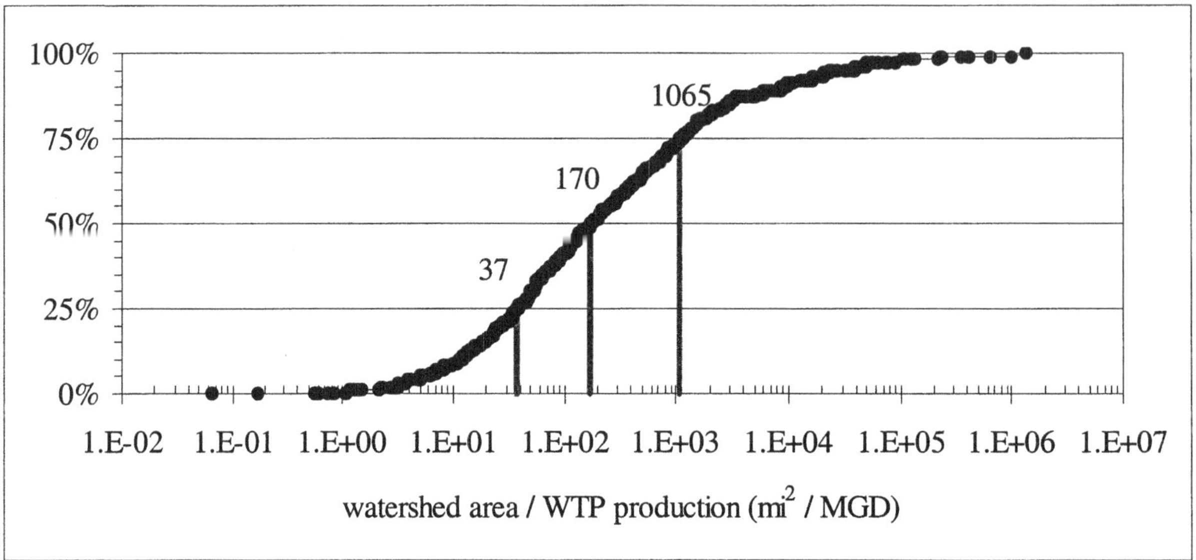

Figure 7.1 Frequency distribution of watershed area: WTP production ratio (*Source*: Data from Alexander et al. 1999 and Alexander 2001)

The catchment size: WTP size ratio is used to provide a screening-level estimate of the percent of WTPs that would benefit from BMP implementation. If the break-even ratio for a given BMP and plant response is 3.7 mi^2/mgd, for example, then the frequency distribution indicates that roughly 4% of similar WTPs would benefit from funding 100% of BMP implementation by themselves. Since most watershed protection efforts involve multiple partners and funding sources, it is realistic to consider cost-sharing scenarios. In the above example, the fraction of WTPs benefiting rises to 25% under the assumption that WTPs bear 10% of the total cost of BMP implementation, and that 90% of the cost is shared among state and federal funding agencies and farmers.

COST-BENEFIT ANALYSIS RESULTS

BMPs were selected for cost-benefit analysis based on their ability to provide meaningful reductions in total watershed export of TOC and TSS. Because urban land use was a small source of total pollutant loads in the model watersheds, urban structural BMPs were observed to provide negligible reductions in total pollutant export. Non-structural BMPs such as impervious cover reduction through alternative development practices may be an effective long-term strategy for mitigating the effects of urbanization, but the cost of these practices is poorly documented. Implementation relies in large part on political efforts to establish zoning ordinances.

BMPs that reduce primarily phosphorus loads (e.g. tertiary treatment at WWTPs or nutrient management systems for cropland) were not included in the cost-benefit analysis due to the limitations of the modeling tools in translating these phosphorus reductions into measurable treatment plant impacts. Finally, BMPs such as wetlands and settling basins were not included in the cost-benefit analysis. These BMPs are typically targeted to significant pollutant source areas in a watershed, a scenario that is not adequately characterized by the watershed model.

Based on these criteria, two agricultural BMPs were selected for cost-benefit analysis: riparian buffers and conservation tillage. Both of these BMPs operate at a scale captured by an export coefficient-based watershed model, and both are relatively widely promoted, funded, and implemented. Table 7.1 below summarizes cost-benefit analysis results for these BMPs. All scenarios assume that the utility funds 10% of the total cost of BMP implementation, and that 90% of the cost is shared among state and federal funding agencies, farmers, and other watershed partners.

Table 7.1
Summary of cost-benefit analysis results

BMP	Reduction in TSS (%) TOC (%)	WTP	BMP cost ($/mi^2 watershed)	Treatmen t savings ($/mgd)	break even ratio with 10% cost share (mi^2/mgd)	WTPs benefiting (%)
Riparian buffer treating 35% of cropland ($140/acre/yr)	25% TSS 15% TOC	Corbalis	$1,797	$1,046	5.8	6%
		Millwood	$1,797	$832	4.6	5%
		Muncie	$1,797	$985	5.5	6%
		Otisco	$1,797	$558	3.1	3%
Riparian buffer treating 35% of cropland ($50/acre/yr)	25% TSS 15% TOC	Corbalis	$642	$1,046	16.3	14%
		Millwood	$642	$832	13.0	12%
		Muncie	$642	$985	15.3	14%
		Otisco	$642	$558	8.7	8%
no-till cultivation on 57% of cropland ($25/acre/yr)	40% TSS 20% TOC	Corbalis	$11	$1,572	1,485.0	79%
		Millwood	$11	$1,007	951.4	73%
		Muncie	$11	$1,568	1,481.0	79%
		Otisco	$11	$777	733.5	69%
no-till cultivation on 14% of cropland ($25/acre/yr)	10% TSS 5% TOC	Corbalis	$3	$384	1,451.0	79%
		Millwood	$3	$260	980.3	73%
		Muncie	$3	$345	1,303.0	77%
		Otisco	$3	$1,035	3,911.0	87%

Riparian buffers on prime cropland ($140/acre of buffer/yr) are cost-effective for less than 5% of WTPs similar to those modeled. At lower crop rental rates, however, the practice becomes cost-effective for 8% - 14% of WTPs. Conservation tillage is predicted to be cost-effective for a larger number of WTPs, roughly 70 to 80% under both implementation scenarios.

While the approach presented above offers a screening-level estimate of BMP cost-effectiveness, it is limited in several respects. First, the modeling approach does not account for pollutant transport or fate in receiving waters, and therefore does not describe the effects of pollutant source area or BMP location within the watershed on the magnitude of pollutant loads delivered to the intake. It assumes that two identical pollutant sources will have an equal impact on raw water quality at the intake, regardless of their location in the watershed. As watershed size increases this assumption becomes less accurate. The methodology outlined above thus overestimates the land area that significantly impacts raw water pollutant concentrations, and in so doing underestimates the fraction of WTPs that will find it cost-effective to implement BMPs.

Second, the watershed - WTP size frequency distribution is based on the 60% of WTPs serving over 10,000 people found on reaches large enough to be included in the RF1 data set. It does not include the 40% of WTPs found in smaller watersheds, and therefore likely underestimates BMP cost-effectiveness. It also does not describe the catchment size relationship for utilities serving fewer than 10,000 people.

Finally, since only TOC and TSS are included in the WTP modeling, this method does not quantify the benefits of reduced loads of phosphorus, nitrogen, pesticides, or other parameters. Many agricultural BMPs will reduce export of constituents other than organic carbon and sediment, and will therefore achieve potentially significant water quality benefits and treatment cost reductions that are not accounted for with this method. Overall the method is conservative in its estimation of BMP cost-effectiveness.

CHAPTER 8
PHOSPHORUS

INTRODUCTION

Unlike many contaminants, phosphorus is not a direct health-related concern for drinking water utilities. Instead, it is the effect phosphorus has on the source of supply that is of most concern. Excessive phosphorus inputs to receiving waters accelerate the natural process of eutrophication and lead to a variety of operational and aesthetic water treatment problems. In addition, phosphorus has an indirect effect on public health by increasing autochthonous production of DBP precursors.

PHOSPHORUS LOADS AND EUTROPHICATION

Phosphorus loads increased from the 1950s through the 1970s as a result of increased manure production and use of fertilizer and detergents. In response to well-publicized eutrophication problems (e.g. Lake Erie), however, point source phosphorus loads have since decreased. These decreases resulted in large part from the use of secondary treatment in wastewater treatment plants pursuant to the 1972 Federal Water Pollution Control Act (Clean Water Act), partial or complete phosphate laundry detergent bans in 27 states, and the detergent industry's phase-out of phosphates in laundry detergents by 1994. These actions have helped reduce wastewater effluent phosphorus concentrations from roughly 11 mg/L in 1970 to 5 mg/L today. Wastewater phosphorus effluent limits in some areas have forced the implementation of tertiary treatment, which has further reduced effluent phosphorus concentrations to 0.5 – 1.5 mg/L (Litke 1999).

As point source phosphorus loads have decreased, attention has come to focus on nonpoint source loads. USGS national regression model (SPARROW) estimates of phosphorus sources indicate that point sources are on average a relatively small contributor to total phosphorus exports. Based on predictions for the 2052 eight-digit hydrologic cataloging units (HUCs) in the conterminous U.S., point sources contribute on average only 11% of total phosphorus export within a given HUC, while nonpoint sources including fertilizer (20%), livestock waste (29%), and nonagricultural nonpoint sources (40%) are the most significant sources of phosphorus loads (based on data described in Smith et al. 1997). The mineral component of the soil transported during runoff events due to hydrologic modification is an important source of phosphorus that has not been well quantified.

Due in large part to nonpoint sources of phosphorus, eutrophic conditions persist in many surface waters. Of the 32 states that assessed lake trophic status in the 1998 National Water Quality Inventory, 17% of assessed lakes were oligotrophic, 33% were mesotrophic, 38% were eutrophic, and 12% were hypereutrophic (USEPA 2000b). USGS NAWQA Phase I results indicate that the nation's streams and rivers frequently exceed the EPA target level of 0.1 mg/L phosphorus for preventing nuisance plant growth. 57% of all streams sampled, and about 75 percent of streams in agricultural and urban areas, exceeded the 0.1 mg/L criterion (USGS 1999).

EFFECTS OF PHOSPHORUS ON RAW WATER QUALITY AND TREATMENT

Excess loads of TP on receiving waters accelerate the eutrophication process and lead to excessive plant growth (see e.g. Cooke and Carlson 1989). Plant growth may be classified as phytoplankton (freely floating algal cells, clumps or mats), periphyton (algae and other microorganisms that attach to substrates), and macrophytes (floating or rooted plants) (USEPA 1999b). When this plant growth dies it settles and decays, consuming dissolved oxygen. Negative impacts of eutrophication on water treatment plants include problems triggered by biomass growth and decay:

- *Filter clogging:* Algae have a low density and do not settle well in conventional treatment plants. Excessive raw water algae concentrations can cause filter clogging, which leads to reduced filter run times, increased backwash frequency, increased energy consumption, and reduced plant output. In order to mitigate these adverse effects, waters with high algae loads are often treated with dissolved air flotation.
- *pH fluctuations:* When algae photosynthesize during the day they consume CO_2, which increases raw water pH, and when they respire at night they produce CO_2, which reduces pH. Algae blooms can cause severe swings in raw water pH; fluctuations of more than an order of magnitude are not uncommon (from the mid to high 7s to low 9s). Rapid changes in pH make control of process pH and coagulation more difficult and increase acid/base consumption. If pre-pH adjustment (acid addition) is not practiced then filtered water particle counts typically increase and decrease on a diurnal cycle that corresponds to the pH swings.
- *Increased NOM:* The growth and decay of algae and other aquatic plants can be an important autochthonous (internal) source of NOM to raw waters (Cooke and Carlson 1989). This increased NOM concentration affects coagulation and filtration, increases DBP formation potential, and may increase DBPs if sufficient TOC is not removed during treatment.
- *Taste and odor:* Decaying vegetation and algal or bacterial metabolites are the most frequent causes of unpleasant tastes and odors in drinking water (Mallevialle and Suffet 1987). Actinomycetes bacteria and blue-green algae (cyanobacteria), for example, are common in eutrophic waters. Both produce several taste and odor-causing compounds, including geosmin and MIB (methylisoborneol). Biomass decay reduces dissolved oxygen concentrations and can lead to anaerobic conditions at the bottom of the water body, especially in stratified temperate lakes and reservoirs. Anaerobic conditions may allow the release of hydrogen sulfide gas, which causes unpleasant odors. These taste and odor problems lead to increased use of oxidants and/or powdered or granular activated carbon and often result in increased customer complaints.
- *Iron and manganese:* Anaerobic conditions also allow the release of reduced (soluble) iron and manganese from sediments. This typically causes an increase in the use of oxidants at the treatment plant. In addition, if the iron and manganese are not adequately removed, deposits or stains on plumbing fixtures often result and may trigger an increase in customer complaints.

The eutrophication-related treatment problems cited above will increase operating treatment costs for water utilities. Eutrophication of source waters may also require treatment processes changes such as ozonation, installation of GAC filter caps, installation of acid feed or PAC feed facilities, and conversion from direct filtration to conventional treatment or conventional clarification to dissolved air flotation.

EUTROPHICATION MODELS

In order to quantify the effects of eutrophic waters on treatment plants, one must predict the raw water quality changes that accompany excess TP loads. Numerous eutrophication models of varying degrees of complexity have been developed to this end. USEPA (1997a) provides a useful overview of watershed and receiving water models, including examples and contact information. Reckhow and Chapra (1999) offer a brief history of mechanistic and empirical eutrophication models, and describe recent scientific and mathematical modeling advances that have improved predictive capabilities. Koelmans et al. (2001) provide a brief review of general features of eutrophication models in the context of more sophisticated models.

Early empirical models include Sawyer's (1947) recommended nutrient thresholds for preventing algal growth in temperate lakes and Vollenweider's (1968) predictions of trophic status as a function of nutrient load, lake depth, and residence time. While these empirical models predict nutrient concentrations or trophic levels, they do not predict parameters with a direct impact on water treatment (e.g. TOC). Current mechanistic models (e.g. QUAL2E, WASP5, CE-QUAL-W2) suffer the same limitation. These models provide integrated simulation of water body hydrodynamics, nutrient cycling, and simple food-chain development. Parameters modeled typically include nutrient concentrations, BOD, dissolved oxygen, pH, and conservative elements. Chlorophyll-a is commonly measured and modeled as a surrogate for planktonic algal biomass.

Prospects for predicting raw water TOC concentrations are improving. Reckhow and Chapra (1999) describe three advances in mechanistic models, including improved hydrodynamic simulation, more sophisticated simulation of sediment oxygen demand and nutrient release, and explicit modeling of carbon (see e.g. Connolly and Coffin, 1995). Improved modeling of carbon is important for several reasons. Chlorophyll-a: carbon ratios vary significantly depending on nutrient and light availability, so chlorophyll-a is not a reliable surrogate for algal biomass organic carbon concentration. Furthermore, organic carbon modeling is required to more accurately predict the fate and transport of organic pollutants. Koelmans et al. (2001), for example, describe five models that integrate descriptions of eutrophication, contaminant fate, food web dynamics, and food chain bioaccumulation. As the number of processes increases, however, so does the number of parameters required for model calibration, so use of sophisticated models requires significant expertise and budget.

Chlorophyll-a concentration frequently correlates with total P concentration, and site-specific empirical models have been developed for predicting the frequency of nuisance algae levels as a function of TP concentration (Walker and Havens, 1995). Algae blooms are typically considered a recreational nuisance when chlorophyll-a concentrations exceed 20 – 60 ppb (see e.g. Heiskary and Walker, 1995). Based on data for 17 small southeastern drinking water reservoirs, Raschke (1994) recommended a mean growing season limit of less than or equal to 15 ppb chlorophyll-a.

Walker (1983) addresses the impact of reservoir eutrophication on organic carbon concentration and its subsequent impacts on drinking water treatment processes, including DBP formation. Based on data from 37 lakes and reservoirs, Walker observed that median total phosphorus concentration explained 85% of the variation in median TOC concentrations, and offered the following relationship:

$$TOC = 0.56 * TP^{0.63} \quad (TOC \text{ in mg/L, } TP \text{ in } \mu g/L)$$

This relationship predicts that increases in TP load will increase TOC concentration, though the magnitude of this effect decreases as TP levels increase. That is, a 20% increase in TP load will increase TOC concentration more in an oligotrophic lake than in a eutrophic lake.

These relationships demonstrate that increases in phosphorus loadings increase raw water TOC levels. In addition, several researchers have shown that there are direct correlations between TOC and THM precursor concentrations (Edzwald et al. 1985; Amy et al. 1987; and Amy et al. 1990). It follows then, that increases in phosphorus can lead to increases in THM precursor concentrations. Such increases in precursor concentrations can lead to considerable operational and capital cost increases as utilities meet the expected requirements of the Stage 2 Disinfectant/Disinfection Byproducts Rule. Many utilities will have to improve the removal of precursors by adding additional coagulant, installing GAC filters, or installing membrane filtration. Other utilities will switch to chloramines as a secondary disinfectant.

CHAPTER 9
CRYPTOSPORIDIUM

INTRODUCTION

Cryptosporidium is a genus of protozoan parasite that infects a wide variety of fish, amphibians, reptiles, birds, and mammals. Of the ten named species within the genus, only *Cryptosporidium parvum* is considered a threat to humans. The life cycle of *Cryptosporidium* involves multiple stages in the intestinal tract of a human or animal host. Transmission occurs when the infective oocyst is shed from the host during defecation and is later ingested by another host. Oocysts are resistant to adverse environmental conditions and can remain viable in the environment for months.

C. parvum has been recognized as a human pathogen since 1976 when the first case of cryptosporidiosis was diagnosed (CDC 1995; Craun et al. 1998). Since then, several waterborne outbreaks have occurred, including the 1993 Milwaukee outbreak that caused illness in over 400,000 people and death in at least 50 (USEPA 2001e).

Cryptosporidium is a significant contaminant of concern because it is widespread in the environment, has caused several waterborne outbreaks, and is resistant to chlorination. In order to improve control of *Cryptosporidium*, and microbial pathogens in general, the USEPA (USEPA 2001e) is proposing the Long Term 2 Enhanced Surface Water Treatment Rule (LT2ESWTR). The new rule will contain provisions for 1) source water monitoring for *Cryptosporidium*, 2) additional *Cryptosporidium* treatment for filtered systems based on source water *Cryptosporidium* concentrations, 3) inactivation of *Cryptosporidium* by all unfiltered systems; 4) disinfection profiling and benchmarking to assure continued levels of microbial protection while public water systems take steps to comply with new disinfection byproduct regulations, 5) covering, treating, or implementing a risk management plan for uncovered finished water reservoirs, and 6) criteria for a number of treatment and management options (the "microbial toolbox") which may be implemented to meet additional *Cryptosporidium* treatment requirements. The USEPA believes that implementation of the LT2ESWTR would significantly reduce levels of *Cryptosporidium* in finished drinking water, which would reduce the rates of endemic cryptosporidiosis.

In order to comply with this rule, many utilities will have to install advanced treatment techniques or implement other control options from the microbial toolbox depending on their source water *Cryptosporidium* concentrations. Although several treatment options are available, minimizing filtered water turbidity (to less than 0.1 NTU 95% of the time), installing ultraviolet light disinfection or ozonation, and implementation of a watershed protection program are likely alternatives for many systems. The effectiveness and cost for each alternative is site-specific.

OCCURRENCE OF CRYPTOSPORIDIUM

Raw water *Cryptosporidium* oocyst concentrations were sampled in large and medium water systems as part of the USEPA Information Collection Rule (ICR) and the Information Collection Rule Supplemental Study (ICRSS). The methodology used during the ICRSS was an improvement over the ICR method. The ICRSS data showed that 97% of water intakes located on flowing streams had at least one sample (out of 24 samples over a 1 year period) in which

Cryptosporidium was detected, while 78% of intakes located in reservoirs or lakes had at least one positive sample. The median raw water oocyst concentration in both the ICR and ICRSS was roughly 0.05 oocysts/L (USEPA 2001e). LeChevallier and Norton (1995) presented data from 262 raw surface water samples and 262 finished water samples collected from 72 surface water supplies from 1991 to 1993. *Cryptosporidium* was found in 52% of the raw water samples and 13% of the plant effluent samples (although many of the oocysts were non-viable). The geometric mean concentration of detectable oocysts in the raw water was 2.1 oocysts/L, with levels ranging from 0.065 to 65.1 oocysts/L. The geometric mean concentration in the finished water was 3.3 oocysts/100L. These data indicate that the concentration of oocysts in a raw water supply can be highly variable.

SOURCES OF CRYPTOSPORIDIUM

Although a wide variety of animals are reservoirs for *Cryptosporidium* oocysts, cattle, pigs, and cats exhibit evidence for transmitting the disease to humans (Rose 1997). Young dairy calves in particular have been observed to shed large numbers of oocysts: Nydam et al. (2001) estimated that an infected calf would shed on the order of 10^{10} oocysts in its first twelve days of life.

Sischo et al. (2000) studied the prevalence and risk factors for shedding of *Cryptosporidium* by dairy cattle and calves and the prevalence and risk factors for *Cryptosporidium* occurrence in surface waters associated with dairy farms in a well-defined watershed in the northeastern United States. Eleven dairy farms were sampled on a monthly basis over a 6-month period. Ninety-one percent of the dairy farms studied had *Cryptosporidium* on their premises. The risk factors for calves shedding *Cryptosporidium* were determined to be contact between calves and frequent bedding changes. The probability of shedding *Cryptosporidium* decreased with age (see also Garber et al. (1994), Ongerth and Stibbs (1989), Nydam et al. (2001)). Even though 91% of the farms had *Cryptosporidium* on the premises, only nine percent of farm-associated stream samples were *Cryptosporidium*-positive. The researchers found that the single risk factor for detecting *Cryptosporidium* in surface water was increased frequency of spreading of manure on fields. The probability of detecting *Cryptosporidium* in streams decreased as 5-day cumulative precipitation increased. There were no animal-associated or barnyard-management features associated with detecting *Cryptosporidium* oocysts in farm-impacted streams.

Graczyk et al. (2000) studied environmental and geographical factors contributing to watershed contamination with *Cryptosporidium*. They examined cattle farms in the floodplain area of Lancaster County, Pennsylvania and found that, on average, a single farm comprised 12.8 hectares of farmland at risk of inundation. In all farms studied, cattle had unlimited access to the creek. Manure samples collected from closed-in calf pens, cow/heifer yard runoff, and cattle paths through the creek were tested for *Cryptosporidium*. On 64% of the farms (n=50) at least one sample tested positive for *Cryptosporidium* and 44% of the farms had oocysts in all manure samples. The researchers point out that the migration of *Cryptosporidium* through soil is insignificant, and that drainage from manure storage areas, direct contact by cows with water, and runoff from fields on which manure has been spread are the most likely modes by which *Cryptosporidium* is transported to surface waters. Suggested reduction strategies include restricting the access of cows to streams by the installation of fencing, containment and decomposition of manure in lagoons, and disinfection of raw manure stored near water.

Sewage treatment plants can also be significant sources of *Cryptosporidium* oocysts. Concentrations of oocysts in untreated wastewater have been observed to range from 1 to 120 oocysts/L (Rose 1997). Factors affecting oocyst concentrations in untreated wastewater include the prevalence and intensity of human and animal infection within the catchment and the contribution of animal waste to the sewage (Robertson et al. 1999). Oocyst removal through wastewater treatment processes varies. Rose et al. (1996) observed a 3.27 log reduction in raw wastewater oocyst concentrations through a tertiary treatment plant. Unit processes and removal performance included activated sludge and clarification (1.14 log removal), coagulant and polymer addition followed by granular media filtration (1.68 log removal), chlorination (0.41 log removal), and storage (0.04 log removal). *Cryptosporidium* detection frequency ranged from 67% in raw wastewater to 17% in the storage tank, at maximum concentrations ranging from 120 oocysts/L (raw) to 0.05 oocysts/L (post-chlorination). Rose et al. (2001) later observed 4.6 log protozoa reduction at an advanced wastewater reclamation plant providing conventional primary and secondary treatment, high pH chemical treatment, two-stage recarbonation with intermediate settling, multimedia filtration, and chlorination. Advanced wastewater treatment processes can thus reduce effluent oocyst concentrations to levels equal to or less than that of typical ambient surface waters.

ENVIRONMENTAL EFFECTS ON *CRYPTOSPORIDIUM* OOCYSTS

The effect of environmental conditions on *Cryptosporidium* oocyst viability was investigated in lab studies by Sattar et al. (1999). Freeze-thaw cycling, shearing, and desiccation were observed to have a negative effect on oocyst viability, as well as exposure to sunlight, temperatures greater than 25°C, and microbial antagonism. Anaerobic conditions did not have a profound effect on survivability: roughly 20% and 40% of oocysts were viable after 30 days exposure to anaerobic conditions at 4° and 20°C, respectively. More research is merited to assess oocyst viability under typical cow and swine manure lagoon conditions. Freeze-thaw cycling was observed to decrease viability substantially: three freeze-thaw cycles at -20°C generally provided 2 log inactivation, suggesting that winter manure stockpiling is beneficial.

This observation was further expanded by Jenkins et al. (1999), who studied the effect of typical manure management practices (manure piles and land application of manure) on oocyst survivability at three dairy farms in the New York City watershed. Small-volume sentinel chambers developed to equilibrate with external chemical and moisture conditions were filled with soil or waste from dairy calves, inoculated with oocysts, and then placed in waste piles or in soil in a cropped field on the third farm for a winter season and a fall-winter season. Infective oocysts in soil and manure decreased to less than 1%. Reduced viability was attributed to the combined effects of temperature cycling and possibly the physical effects of abrasion by soil particles. This research suggests that fall stockpiling or land application of manure can achieve roughly 2 log oocyst inactivation by the following spring.

Olson et al. (1999) observed that cold temperatures were the most favorable conditions for oocyst survival. Researchers measured *Giardia* cyst and *Cryptosporidium* oocyst viability in water, cattle feces, and soil at temperatures of -4, 4, and 25°C over a period of 12 weeks. *Cryptosporidium* oocysts survived greater than 12 weeks in water and feces at -4° and 4°C. Survival was lowest at 25°C. *Giardia* cysts in water, feces, and soil were inactivated within 2 weeks at -4° and 25°C, but lasted up to 7 weeks in soil and 11 weeks in water at 4°C. Based on these observations, manure should be stored for at least 12 weeks and land applied during warmer months. While this strategy is viable on most grassland, it is more difficult to implement on cropland, where the presence of a growing crop impedes manure application.

SOURCE WATER PROTECTION FOR CONTROL OF *CRYPTOSPORIDIUM*

The export of *Cryptosporidium* from watersheds and its transport and fate in receiving waters is highly site-specific, so no quantitative *Cryptosporidium* modeling was conducted for this study. On a qualitative level, watershed protection measures are a conceptually sound means of protecting against *Cryptosporidium*. The multiple barrier approach has long been accepted as a core principle of drinking water treatment, and reducing export of *Cryptosporidium* oocysts to receiving waters is an important first barrier. This philosophy is embodied in the expected terms of the LT2ESWTR, which will allow water treatment plants with raw water oocyst concentrations greater than 0.075 oocyst/L to meet some of their additional treatment requirements through watershed protection measures.

Practically speaking, however, the effectiveness of BMPs in reducing *Cryptosporidium* export is still poorly quantified. The effect of practices such as improved animal waste management and buffer strips on *Cryptosporidium* export from intensive agricultural watersheds merits further study. Further research is also required to characterize BMP performance and reliability during extreme precipitation events.

At wastewater treatment plants, studies have shown that *Cryptosporidium* removal through advanced processes is significantly greater than removal through conventional secondary treatment. The impact of advanced process addition at conventional WWTPs on total *Cryptosporidium* load reductions should be assessed relative to the oocyst contribution from livestock and wildlife sources.

CHAPTER 10
CONCLUSIONS

INTRODUCTION

The first section of this chapter summarizes project findings regarding the effects of point and nonpoint source pollutant loads on water treatment plants, the potential of BMPs to reduce pollutant loads to source waters, the impact of pollutant loads on treatment costs, and the cost-effectiveness of BMPs as a means of reducing treatment costs. Section two discusses the impact of modeling methodology limitations on project findings, and section three summarizes research needs identified in this project.

CONCLUSIONS

Major Point and Nonpoint Pollutant Sources

Export coefficient-based watershed modeling was conducted to investigate the relative contribution of point sources and nonpoint sources to total watershed export of TOC, TP, and TSS. Modeling scenarios were based on hypothetical watersheds with land use and population density believed to be characteristic of typical forested, agricultural, and mixed land use drinking water supply watersheds. Major findings for these watersheds include:

- The majority of TOC, TP, and TSS exported to receiving waters comes from nonpoint sources. The relative contribution of different nonpoint sources depends heavily on the fraction of total watershed area devoted to a given land use. In forested watersheds, forest cover is the dominant source of TOC and a major source of TP and TSS. In agricultural watersheds, cropland is the dominant source of TOC, TP, and TSS. Pollutant contributions in mixed land use watersheds depend in large part on land use and population density. Forest and cropland are major sources of TOC and TSS. Cropland and point sources (i.e. wastewater treatment plants) are the dominant sources of TP.
- The primary water quality impact of pasture and grazing land is associated not with surface runoff but with direct livestock access to streams and stream banks, which increases sediment, nutrient, and pathogen loads to receiving waters.
- Runoff from concentrated animal feeding operations depends on site-specific waste-handling systems. Most CAFO waste is ultimately disposed via land application, so the impact of CAFOs on pollutant loads depends largely on land application practices (e.g. cropland available for land application, application rates, timing, etc.). Land application of CAFO waste at non-agronomic rates is common and leads to phosphorus buildup in soils and increased phosphorus leaching and runoff potential. Model predictions indicate that CAFOs of all types are expected to make significant contributions to TOC and TP loads in areas where manure storage and treatment is inadequate and manure application rates exceed crop utilization rates.

- For the watersheds modeled in this project, pollution from urban land use alone, even at relatively high population densities, is expected to be a small fraction (<10%) of the total watershed export of TOC, TP, and TSS. For the pollutants considered in this project, the major impact of population growth and development is increased TP loads from wastewater treatment plants and increased channel erosion.
- Increased impervious cover from population growth and development contributes to channel erosion. Though the magnitude of the pollutant loads resulting from channel erosion are highly site-specific, modeling results indicate that channel erosion resulting from moderate levels of development can be a major contributor to total watershed export of TOC, TP, and TSS.
- Increased wastewater treatment plant discharges from population growth can significantly increase total watershed export of TP, even in low population density watersheds.
- Pollutant loads from failing septic systems, CSOs, and SSOs are highly site-specific and difficult to predict. Model predictions using best available load estimation parameters indicate that these sources are not typically major contributors to total annual export of TOC, TP, and TSS.

Potential of BMPs to Reduce Pollutant Loads

Export coefficient-based watershed modeling was conducted to investigate the potential of BMPs to reduce total watershed export of TOC, TP, and TSS. Modeling scenarios were based on typical pollutant removal efficiencies for common urban and agricultural BMPs. Findings regarding the potential of BMPs to reduce pollutant loads include:

- Agricultural BMPs such as buffer strips and conservation tillage can reduce the export of TP, TSS, and pesticides from cropland, though the magnitude of the reduction depends heavily on the fraction of total cropland treated by BMPs. The impact of agricultural BMPs on TOC export is not well characterized and merits further research.
- Improved animal waste management at CAFOs and pasture/range BMPs such as livestock exclusion are expected to reduce export of TOC, TP, and TSS in watersheds with intensive animal production.
- Natural forest cover and wetlands are significant sources of TOC and DBP precursor material, and reducing the contribution of these sources with BMPs is not expected to be feasible.
- Because urban land use is expected to contribute a small fraction of total TOC, TP, and TSS export in most drinking water supply watersheds, urban BMPs have limited potential to reduce total export of these parameters. Load reduction potential is further reduced by practical implementation limitations. Urban BMPs are expected to provide significant reductions of TOC, TP, and TSS only for WTPs treating source waters heavily impacted by highly urbanized subwatersheds.
- Reducing the impervious fraction of new development is a sound strategy for mitigating the long-term effects of population growth and land development,

although the cumulative effect of this technique on sprawl should be considered in the comprehensive watershed planning process. Focused densities also play a role in reducing overall watershed imperviousness.

- BMPs are most effective in reducing total pollutant loads when targeted to the most significant pollutant sources in a watershed. Programs that target BMP implementation in priority pollutant source areas will be more effective in reducing total pollutant loads than those that do not.
- The long-term performance of most structural urban and agricultural BMPs depends heavily on effective maintenance.

Impact of Pollutant Loads on Water Treatment Plants

Dynamic water treatment plant modeling was conducted to investigate the effects of changes in raw water TOC and TSS concentrations on WTP operations, costs, and finished water quality. Modeling scenarios were based on four different treatment plant types treating four different source waters. Operating costs quantified include chemical consumption, residuals disposal, and filter backwashing. Findings regarding the impact of pollutant loads on treatment costs include:

- Increases in raw water TOC and TSS concentrations increase WTP operating costs due to increased residuals disposal, filter backwashing, and chemical consumption. Moderate deteriorations in raw water quality (e.g. 25% increase in solids and 25% increase in organics) can increase treatment costs by roughly 10%. Moderate improvements (e.g. 25% decrease in solids and 15% decrease in organics) can reduce treatment costs by roughly 5%. Operating costs do not exhibit high sensitivity to changes in TOC and TSS concentrations (that is, the percent change in operating costs is less than the percent change in pollutant loads). The observed changes in operating costs are nonetheless significant.
- Increases in point and nonpoint source phosphorus loads accelerate the natural process of eutrophication in surface waters, which decreases treatability, increases treatment costs, and may increase concentrations of DBP precursors. The magnitude of these changes is highly site-specific and is difficult to quantify on an average basis.
- Export of agricultural pesticides to receiving waters depends heavily on application methods and timing relative to rainfall events. Raw water atrazine concentrations in excess of the current 3 ppb finished water maximum contaminant level (MCL) are common after spring rainfall events in high atrazine-use watersheds. Treatment plants in these watersheds may incur substantial additional treatment costs in order to comply with water quality standards.
- Under the expected terms of the Long Term 2 Enhanced Surface Water Treatment Rule (LT2ESWTR), WTPs drawing from raw water sources with *Cryptosporidium* concentrations greater than 0.075 oocysts/L will be required to provide additional treatment. Depending on raw water oocyst levels, watershed protection measures may satisfy some or all of these additional treatment requirements. Treatment plants drawing from source waters containing more than

1 oocyst/L are expected to incur substantial costs for the addition of new processes such as ozonation, UV irradiation, or membrane filtration.

BMP Cost-Effectiveness

Findings regarding BMP cost-effectiveness include:

- Targeting BMPs to priority pollutant source areas is critical for maximizing the impact of source protection funds. Many BMP programs do not target BMP funding to the most significant pollutant sources in the watershed.
- BMP cost-effectiveness depends in large part on the scale of the watershed protection effort and the fraction of total BMP implementation cost that is shared by the utility. WTPs would seldom benefit from funding 100% of BMP implementation as a means of reducing routine operating costs. However, most watershed protection efforts involve multiple partners and funding sources. Utilities may be able to improve or protect source water quality and reduce treatment costs at minimal direct expense by forming source water protection partnerships with federal, state, and local interests. Utility participation in watershed protection efforts, even at zero or low BMP cost-share levels, can help prioritize allocation of state and federal funding sources for water supply protection.
- Promotion or cost-sharing of BMPs such as conservation tillage and riparian buffer strips is expected to be a cost-effective strategy for reducing operating costs for some utilities in watersheds heavily influenced by row-crop agriculture. An analysis of national data on the size of watershed size relative to treatment plant size was used to provide a screening-level estimate of the fraction of water treatment plants that would benefit from funding BMP implementation. Assuming that WTPs fund 10% of the total BMP cost, and that other funding sources are available to provide sufficient cost-share funding to achieve BMP implementation, the fraction of WTPs benefiting from BMP implementation ranged from 5 - 15% for riparian buffers to 70 – 80% for conservation tillage.
- Structural urban BMPs are expected to be a cost-effective strategy for reducing operating costs only for WTPs treating source waters with pollutant loads derived primarily from urban NPS pollution. The long-term effects of population growth and increased impervious cover in urbanizing watersheds are best countered at an early stage through promotion of planning and development practices that reduce impervious cover and prevent channel erosion.
- Depending on treatment plant processes and operations, increases in raw water concentrations of DBP precursors, pathogens, and pesticides have the potential to increase risks to public health. Source water protection measures that reduce this risk will have public health benefits that are not adequately expressed in strict economic terms.
- The modeling scenarios in this report describe the effect of changes in TOC and TSS loads on routine operating costs. They do not quantify the effects of other pollutants on WTP costs, nor do they account for the costs of major process changes or source of supply changes. Severe water quality deterioration may

require increasing the capacity of existing processes, adding new unit processes, or developing alternate sources of supply. Due to the high cost of WTP capital improvements relative to watershed BMPs, utilities should consider promoting long-term source water protection measures that have the potential to prevent major process changes.

- Overall, this report is believed to present a conservative assessment of BMP cost-effectiveness.

METHODOLOGY LIMITATIONS

A coupled watershed-water treatment plant modeling approach was selected for this project because it provides a useful framework for assessing the costs and benefits of source water protection measures. Observations regarding the modeling tools applied in this project include:

- The Watershed Treatment Model provides a useful screening-level assessment of the relative importance of different point and nonpoint sources on annual export of TOC, TP, and TSS. It also indicates the magnitude of pollutant load reductions that can be expected with BMP implementation. The model does not simulate fate and transport of contaminants within the watershed, nor does it predict pollutant export from pervious land cover during runoff events.
- Since the model default removal efficiencies for many of the agricultural BMPs are roughly comparable (e.g. 60-80% for solids, 40-60% for TP, and 30-40% for TOC), the load reduction potential predicted by the WTM depends primarily on the fraction of cropland they treat. This is not strictly true in practice, since different land areas have different erosion and pollutant loss potential. Overall load reductions from BMPs depend strongly on where BMPs are situated in the watershed, not just on the percent of land treated.
- In order to provide an assessment of watershed changes on water treatment plants that was not site-specific, pollutant fate and transport in receiving waters was not modeled in this project. Dynamic, site-specific receiving water modeling is required to provide a more accurate account of the effects of pollutant attenuation and transformation in receiving waters on water quality at the intake. This is particularly true for sediment, phosphorus, and organic carbon. More sophisticated receiving water modeling is required to link changes in watershed export of phosphorus with water quality parameters of interest at WTPs.
- The modeling methodology does not account for TOC/DBP precursor production in receiving waters, nor does it account for the variation in DBP formation potential of different TOC sources.
- OTTER is a powerful tool that can effectively simulate most water treatment plant processes and parameters of concern. It has proved useful in this project for assessing the impact of changes in raw water TOC and TSS concentrations on WTP operations, costs, and finished water quality.
- OTTER does not account for some of the significant operating costs of eutrophic source waters such as taste and odor control with PAC and filter clogging due to

algae blooms. The model is also relatively data and labor-intensive and may exceed the resource levels of many small or medium-sized utilities.

- OTTER models are calibrated by adjusting parameters of different unit process in order to match observed and predicted water quality characteristics. Calibration parameters are to a certain extent specific to a given set of raw water quality characteristics. This fact limits the accuracy of model predictions under raw water quality conditions that are radically different from those under which the model was originally calibrated.

RESEARCH NEEDS

The following research needs were identified during the course of this study:

- Better characterization of TOC export and DBP formation potential of different TOC sources is required to improve our understanding of the effects of watershed activities on DBPs in drinking water. TOC is measured less frequently than nutrients and sediment in BMP studies and ambient water quality monitoring, and the effect of most BMPs on the export of TOC and DBP precursors is poorly characterized. This is particularly important for BMPs such as conservation tillage and riparian buffers, which increase plant biomass and soil organic matter carbon levels. These BMPs are widely recommended and implemented; yet their effect on DBP precursor export is poorly understood. Similarly, the export of TOC and DBP precursors from concentrated animal feeding operations using different types of waste handling and disposal systems merits further research.
- Improved modeling and monitoring of the effects of population growth and land development on pollutant loads from channel erosion is necessary to guide source protection in developing watersheds
- Improved modeling and monitoring of the effects of BMPs on *Cryptosporidium* export from agricultural watersheds will enhance the effectiveness of source protection programs implemented under the LT2ESWTR.
- A better national-scale assessment of the size, land use, and population density of water supply watersheds will help guide source water protection policy, funding, and program implementation.

CHAPTER 11
RECOMMENDATIONS TO UTILITIES

Though drinking water treatment has historically been conceived as a series of unit processes at a treatment plant, watershed protection is in a sense the first step in the treatment train. Best management practices in the watershed can help reduce pollutant loads that would otherwise require removal at the plant. In some cases these BMPs may be a cost-effective alternative to treatment at the plant. Determining whether it makes economic sense for a utility to invest in watershed protection as a means of reducing future treatment costs is a complex and site-specific decision-making process influenced by a wide variety of factors. Recommendations for utilities considering source protection include:

- Use the Source Water Assessment Plan to identify significant potential sources of your contaminants of concern within your watershed. When developing a source water protection strategy, target BMP promotion to these areas.
- For utilities in small watersheds, use a simple model such as STEPL or WTM to develop a planning-level estimate of the major pollutant sources in your catchment.
- For all utilities, partner with existing watershed protection or conservation groups in order to emphasize the importance of protecting raw water supplies. Utility participation in watershed protection efforts can help prioritize allocation of BMP funds.
- Start addressing urban NPS pollution now. The effects of population growth are difficult and expensive to mitigate after the fact. Low-cost efforts to reduce increases in new impervious cover through promotion of zoning ordinances and low-impact development practices will help reduce the potential for long-term water quality deterioration.
- Address agricultural NPS pollution from crop and pasture land by partnering with farm groups, county-level Soil and Water Conservation Districts, or extension agents. These stakeholders have considerable knowledge of significant pollutant sources within the watershed, and are well acquainted with BMPs and funding sources. Consider cost-sharing a small percentage of BMP implementation as a means of leveraging additional funding sources and reducing treatment costs.
- Address pollutant loads from CAFOs by promoting improvements in agricultural waste management systems, as well as land application of animal waste at phosphorus-based rates.
- When evaluating investment in source protection measures, consider the long-term impact of these measures on protecting water quality and reducing future treatment costs and facilities requirements. Consider as well the non-economic benefits of source protection such as improved customer confidence and the potential public health benefits of reduced exposure to pesticides, pathogens, and emerging contaminants of concern.

APPENDIX A
WATERSHED MODELING RESULTS

Table A.1
Scenario definitions and load results for urban development runs

run	base case	Urban	Crop	Pasture	Forest	pop/ km^2	WWTP*	Channel Erosion**	BMP 1	% change from base case TOC	TP	TSS
Population growth in a pristine forested watershed												
F1	F1	2%	5%	10%	83%	5	0%	-	-	0%	0%	0%
F2	F1	3%	5%	10%	82%	10	0%	-	-	0%	2%	2%
F3	F1	3%	5%	10%	82%	10	50%	-	-	1%	17%	2%
F4	F1	3%	5%	10%	82%	10	50%	45%	-	6%	53%	40%
F4.1	F4	3%	5%	10%	82%	10	50%	45%	Tertiary WWTP	0%	-7%	0%
F4.2	F4	3%	5%	10%	82%	10	50%	45%	Tertiary WWTP & 20% imperv. Cover reduction	-2%	-14%	-8%
F5	F1	5%	5%	10%	80%	25	80%	76%	-	27%	222%	170%
F5.1	F5	5%	5%	10%	80%	25	80%	76%	Tertiary WWTP	-1%	-13%	0%
F5.2	F5	5%	5%	10%	80%	25	80%	76%	Tertiary WWTP & 20% imperv. Cover reduction	-6%	-27%	-19%
Population growth in an agricultural watershed												
R1	R1	3%	75%	10%	12%	10	0%	-	-	0%	0%	0%
R2	R1	5%	73%	10%	12%	25	60%	-	-	3%	8%	0%
R3	R1	5%	73%	10%	12%	25	60%	56%	-	20%	33%	54%
R3.1	R3	5%	73%	10%	12%	25	60%	-	Tertiary WWTP & 20% imperv. Cover reduction	-14%	-24%	-35%
R4	R1	11%	67%	10%	12%	50	80%	-	-	8%	20%	1%
R5	R1	11%	67%	10%	12%	50	80%	78%	-	54%	89%	148%
R5.1	R5	11%	67%	10%	12%	50	80%	50%	Tertiary WWTP & 20% imperv. Cover reduction	-17%	-30%	-33%
Structural urban BMPs												
M1.0	M1.0	5%	30%	15%	50%	25	100%	-	-	0%	0%	0%
M1.1	M1.0	5%	30%	15%	50%	25	100%	-	Wet ponds treating 50% of urban cover	-1%	-1%	-3%
M1.2	M1.2	16%	30%	15%	39%	75	100%	-	-	0%	0%	0%
M1.3	M1.2	16%	30%	15%	39%	75	100%	-	Wet ponds treating 50% of urban cover	-4%	-2%	-9%
M1.4	M1.4	11%	30%	15%	44%	50	100%	-	-	0%	0%	0%
M1.5	M1.4	11%	30%	15%	44%	50	100%	-	Wet ponds treating 50% of urban cover	3%	1%	4%

* Percent of population served by WWTP

** Percent of impervious cover contributing to channel erosion

Table A.2
Scenario definitions and load results for agricultural runs

run	base case	BMP 1	BMP 2	% implement	% change from base case TOC	TP	TSS	atrazine
A1	A1	-	-	-	0%	0%	0%	0%
A2	A1	buffer strip	-	10%	-4%	-6%	-7%	-9%
A3	A1	buffer strip	-	25%	-10%	-15%	-18%	-23%
A4	A1	buffer strip	-	50%	-20%	-29%	-35%	-45%
A5	A1	buffer strip	-	35.3%	-14%	-21%	-25%	-32%
A6	A1	no till/strip till	-	10%	-3%	-6%	-7%	-4%
A7	A1	no till/strip till	-	14.1%	-5%	-9%	-10%	-6%
A8	A1	no till/strip till	-	25%	-8%	-16%	-18%	-10%
A9	A1	no till/strip till	-	50%	-16%	-31%	-35%	-20%
A10	A1	no till/strip till	-	56.5%	-18%	-36%	-40%	-23%
A11	A1	reduced till	-	10%	-2%	-4%	-5%	-3%
A12	A1	reduced till	-	25%	-6%	-11%	-13%	-8%
A13	A1	reduced till	-	50%	-12%	-22%	-27%	-15%
A14	A1	nutrient management	-	10%	0%	-3%	0%	0%
A15	A1	nutrient management	-	25%	0%	-7%	0%	0%
A16	A1	nutrient management	-	50%	0%	-13%	0%	0%
A17	A1	settling basin	-	10%	-3%	-4%	-5%	0%
A18	A1	settling basin	-	25%	-7%	-9%	-13%	-1%
A19	A1	Wetlands	-	10%	2%	-4%	-8%	-6%
A20	A1	Wetlands	-	25%	4%	-11%	-19%	-15%
A21	A1	integrated pesticide management	-	10%	0%	0%	0%	-3%
A22	A1	integrated pesticide management	-	25%	0%	0%	0%	-6%
A23	A1	integrated pesticide management	-	50%	0%	0%	0%	-13%
A24	A1	no till/strip till	Buffer strip	25%	-14%	-20%	-21%	-24%
A25	A1	buffer strip	Integrated pesticide management	25%	-10%	-15%	-18%	-23%
A26	A1	buffer strip	No till/strip till	25%	-14%	-20%	-21%	-24%
A27	A1	buffer strip	Nutrient management	25%	-10%	-17%	-18%	-23%
A28	A1	nutrient management	No till/strip till	25%	-8%	-18%	-18%	-10%
A29	A1	no till/strip till	-	56.5%	-18%	-36%	-40%	-23%

Table A.3
Scenario definitions and load results for CAFO runs

					% change from base case		
run	base case	CAFO	BMP	% implement	TOC	TP	TSS
A30	A1	1000 AU beef	-	-	13%	16%	1%
A31	A1	1000 AU dairy	-	-	16%	20%	3%
A32	A1	1000 AU swine	-	-	10%	13%	1%
A33	A1	1000 AU poultry	-	-	10%	20%	2%
A34	A1	2000 AU beef	-	-	25%	32%	3%
A35	A1	2000 AU dairy	-	-	31%	40%	6%
A36	A1	2000 AU swine	-	-	21%	27%	2%
A37	A1	2000 AU poultry	-	-	20%	39%	4%
A38	A30	1000 AU beef	Animal waste management	100%	-7%	-11%	-2%
A39	A31	1000 AU dairy	Animal waste management	100%	-8%	-13%	-3%
A40	A32	1000 AU swine	Animal waste management	100%	-6%	-9%	-1%
A41	A33	1000 AU poultry	Animal waste management	100%	-5%	-14%	-2%
A42	A30	1000 AU beef	Wetlands	100%	-15%	-11%	-1%
A43	A31	1000 AU dairy	Wetlands	100%	-18%	-13%	-3%
A44	A32	1000 AU swine	Wetlands	100%	-12%	-9%	-1%
A45	A33	1000 AU poultry	Wetlands	100%	-12%	-13%	-2%

APPENDIX B
WATER TREATMENT PLANT MODELING RESULTS

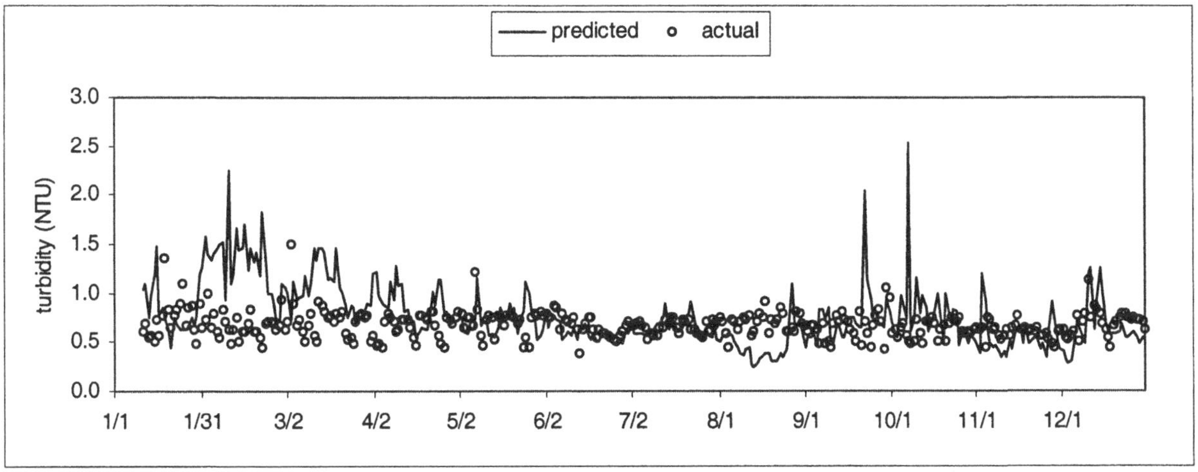

Figure B.1 Corbalis WTP model validation: clarified turbidity

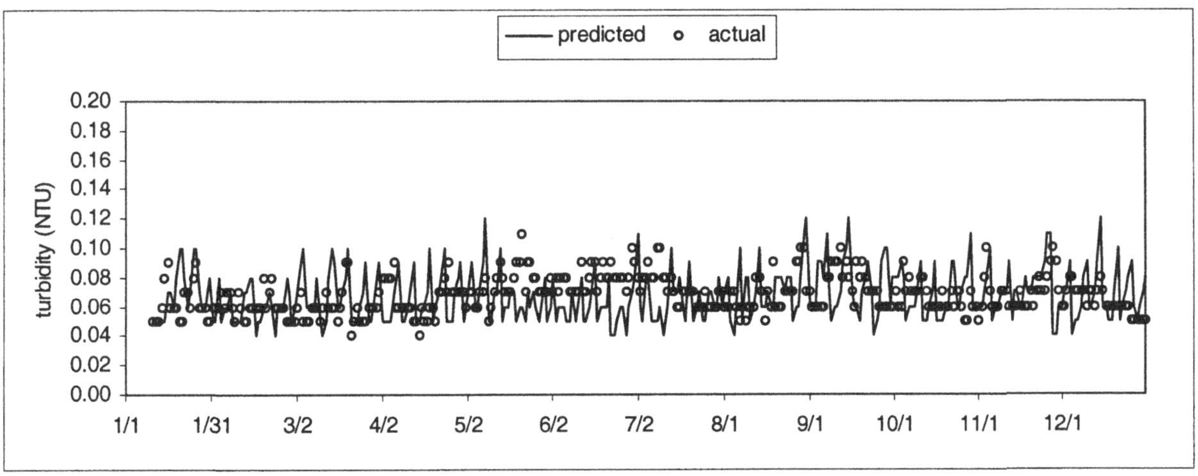

Figure B.2 Corbalis WTP model validation: finished turbidity

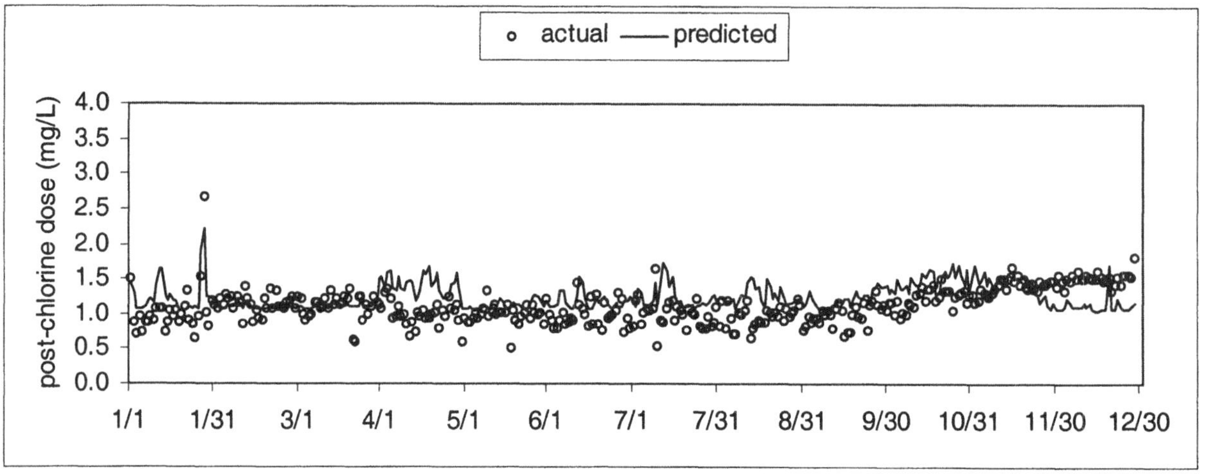

Figure B.3 Millwood WTP model validation: post-chlorine dosage to setpoint

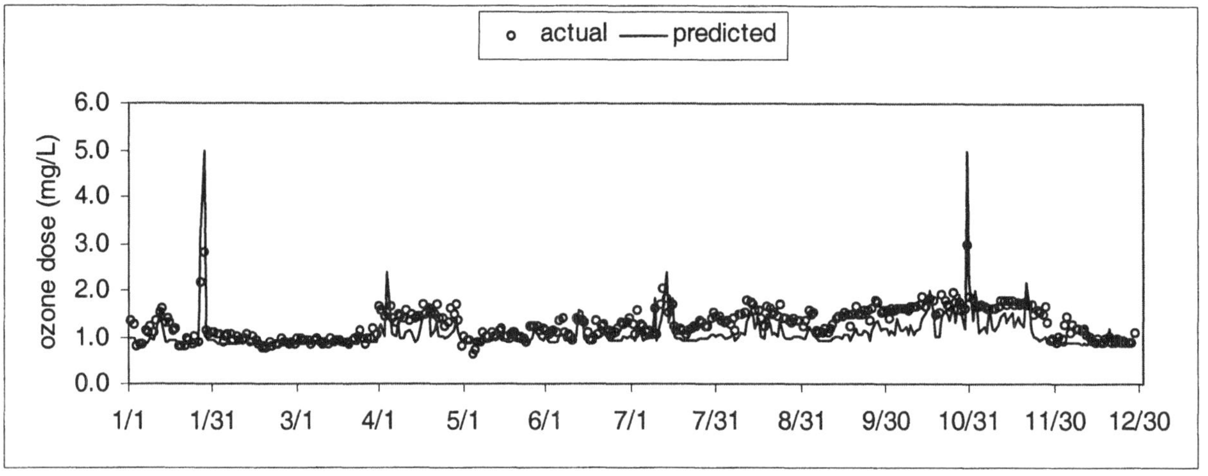

Figure B.4 Millwood WTP model validation: ozone dosage to setpoint

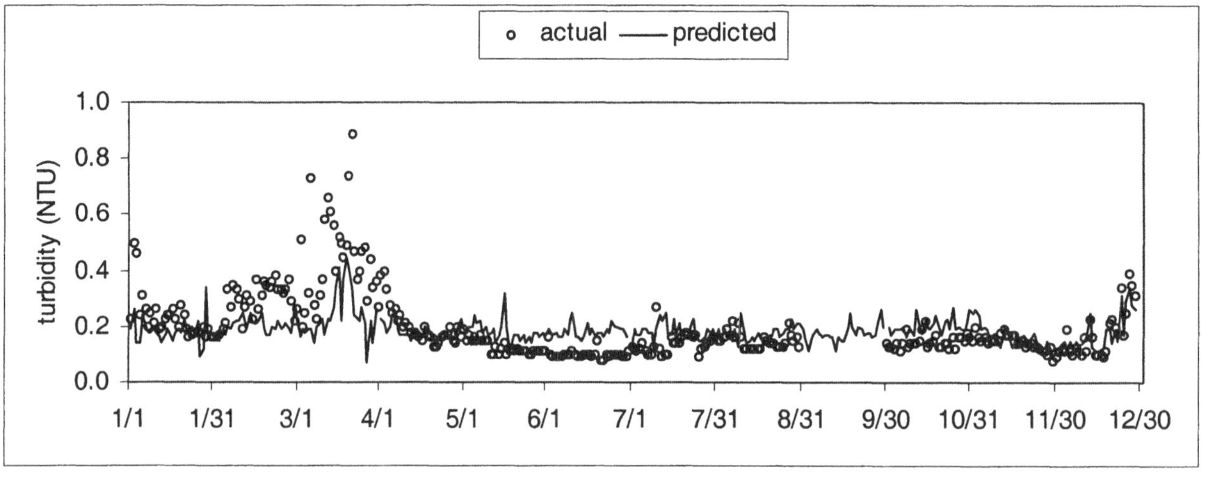

Figure B.5 Millwood WTP model validation: DAF effluent turbidity

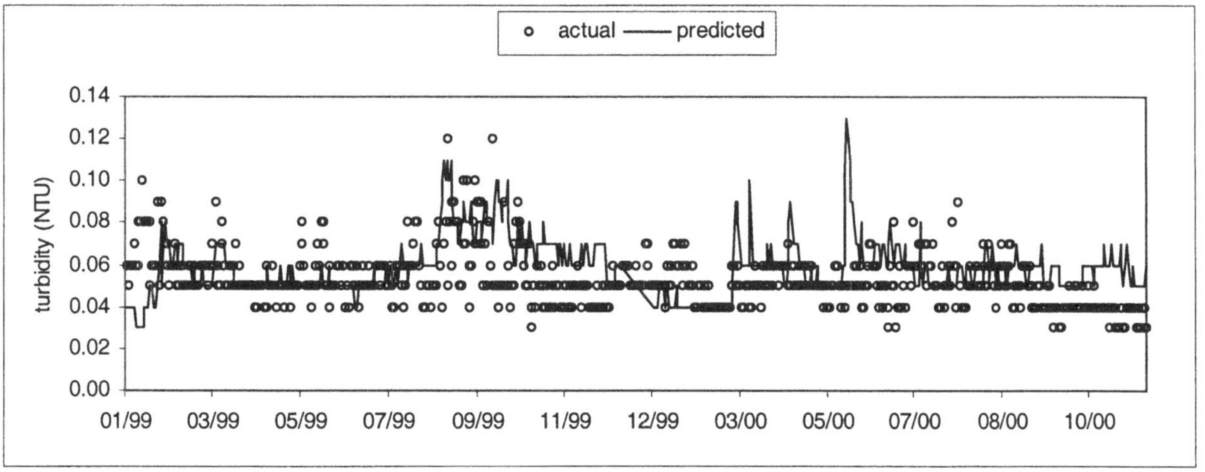

Figure B.6 Otisco WTP model validation: finished turbidity

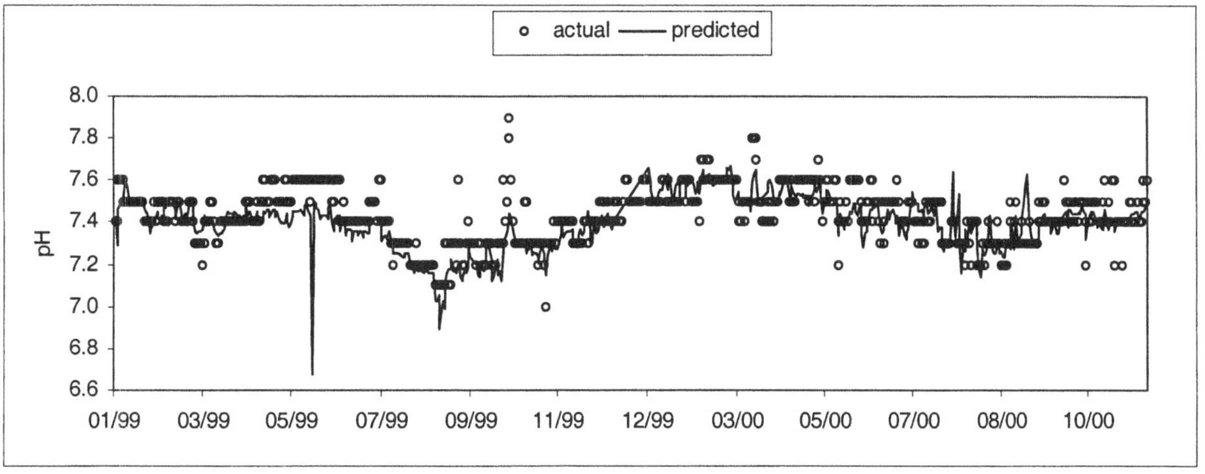

Figure B.7 Otisco WTP model validation: finished pH

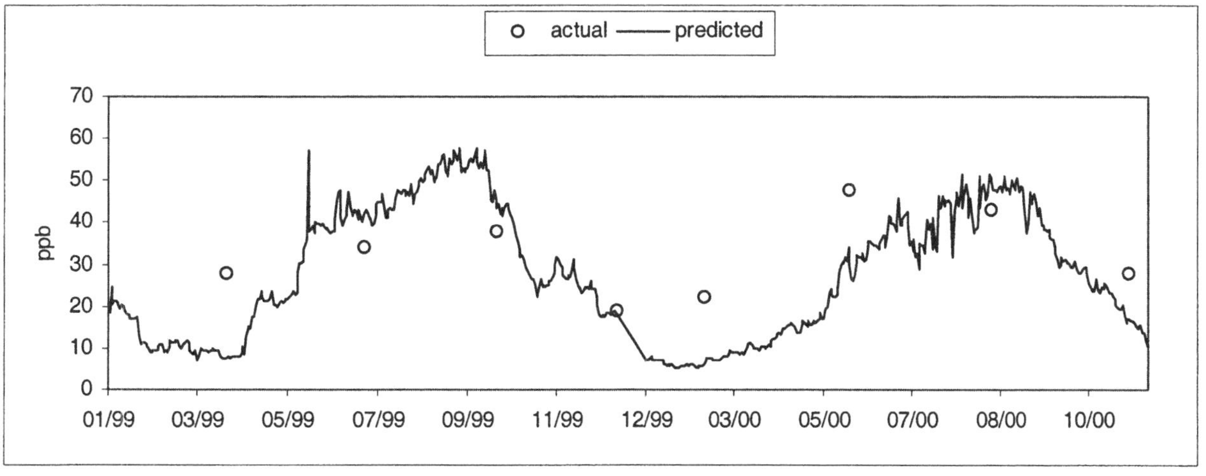

Figure B.8 Otisco WTP model validation: finished TTHM

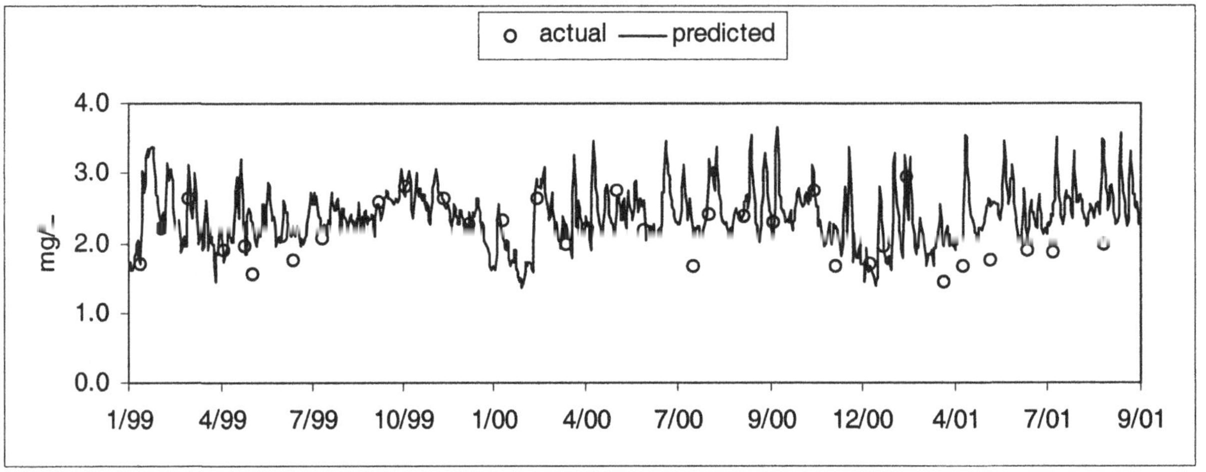

Figure B.9 Muncie WTP model validation: finished TOC

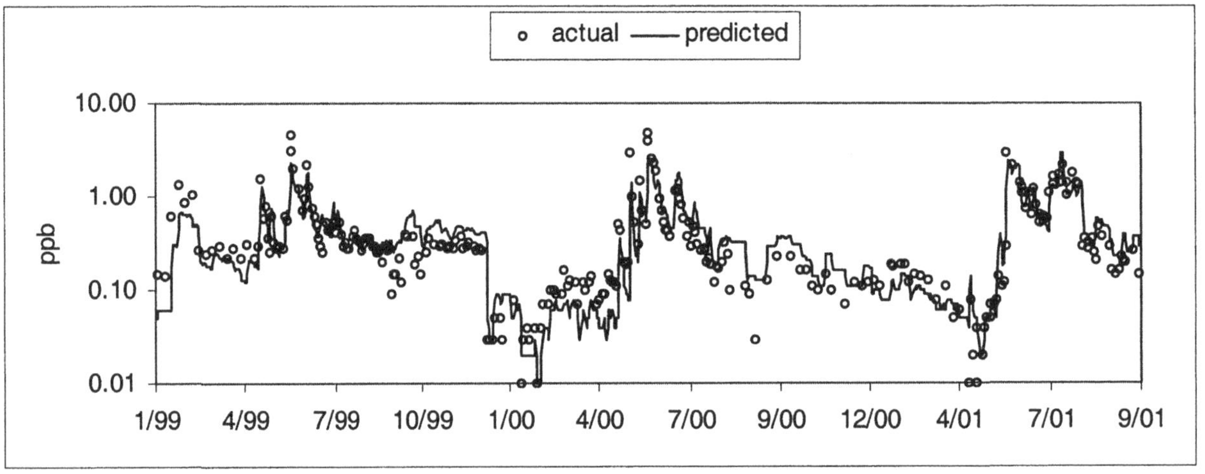

Figure B.10 Muncie WTP model validation: finished atrazine

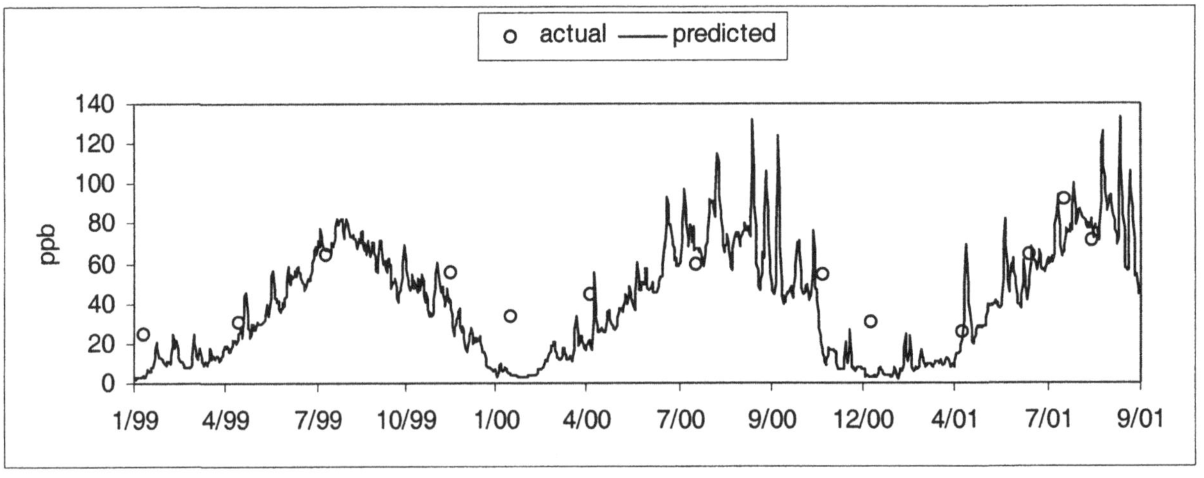

Figure B.11 Muncie WTP model validation: finished TTHM

Table B.1
Summary of WTP modeling results

Scenario	Parameter	Corbalis	Millwood	Muncie	Otisco
1.	average filter applied turbidity	0.66	0.18	1.23	1.68
Base case	average finished turbidity	0.07	0.08	0.03	0.06
	average filter run time (hours)	118	135	53	27
2.	average filter applied turbidity	0.68	0.21	1.23	2.10
	average finished turbidity	0.07	0.09	0.03	0.07
25% increase in solids	average filter run time (hours)	114	117	53	22
	percent change in backwash volume pumped	4%	15%	0%	24%
25% increase in organics	percent change in average finished TTHM	11%	15%	25%	19%
	percent change in average finished HAA5	16%	19%	21%	28%
	percent change in pre-oxidant consumption *	3%	33%	2%	3%
	percent change in coagulant consumption**	7%	2%	9%	17%
	percent change in sludge production	17%	15%	10%	25%
	percent change in post-Cl2 consumption	0%	11%	0%	0%
	percent change in pre-caustic soda consumption	8%	3%	19%	14%
	percent change in post-caustic soda consumption	2%	6%	18%	0%
	percent change in treatment costs	7%	13%	8%	11%
3.	average filter applied turbidity	0.66	0.22	1.27	2.24
	average finished turbidity	0.07	0.09	0.03	0.08
50% increase in solids	average filter run time (hours)	117	115	52	21
	percent change in backwash volume pumped	1%	16%	3%	33%
25% increase in organics	percent change in average finished TTHM	10%	15%	24%	19%
	percent change in average finished HAA5	15%	19%	21%	28%
	percent change in pre-oxidant consumption *	3%	32%	2%	3%
	percent change in coagulant consumption**	9%	4%	11%	19%
	percent change in sludge production	28%	17%	16%	33%
	percent change in post-Cl2 consumption	0%	11%	0%	0%
	percent change in pre-caustic soda consumption	11%	5%	24%	15%
	percent change in post-caustic soda consumption	3%	6%	23%	0%
	percent change in treatment costs	11%	14%	11%	13%
4.	average filter applied turbidity	0.69	0.22	1.34	2.52
	average finished turbidity	0.07	0.09	0.03	0.09
100% increase in solids	average filter run time (hours)	112	112	48	18
	percent change in backwash volume pumped	6%	19%	9%	48%
50% increase in organics	percent change in average finished TTHM	8%	14%	23%	19%
	percent change in average finished HAA5	13%	18%	20%	28%
	percent change in pre-oxidant consumption *	2%	30%	2%	3%
	percent change in coagulant consumption**	15%	7%	14%	22%
	percent change in sludge production	51%	20%	25%	50%
	percent change in post-Cl2 consumption	0%	10%	0%	1%
	percent change in pre-caustic soda consumption	18%	7%	29%	16%
	percent change in post-caustic soda consumption	3%	7%	28%	2%
	percent change in treatment costs	19%	14%	16%	17%

(continued)

Table B.1
Summary of WTP modeling results (continued)

Scenario	Parameter	Corbalis	Millwood	Muncie	Otisco
5.	average filter applied turbidity	0.65	0.18	1.22	1.57
	average finished turbidity	0.06	0.08	0.03	0.05
10% decrease in solids	average filter run time (hours)	119	135	54	30
	percent change in backwash volume pumped	-1%	0%	-1%	-7%
5% decrease in organics	percent change in average finished TTHM	-2%	-3%	12%	-4%
	percent change in average finished HAA5	-3%	-3%	-4%	-6%
	percent change in pre-oxidant consumption *	-1%	-4%	-1%	-1%
	percent change in coagulant consumption**	-2%	-1%	-3%	-4%
	percent change in sludge production	-6%	0%	-4%	-7%
	percent change in post-Cl2 consumption	0%	-2%	0%	0%
	percent change in pre-caustic soda consumption	-2%	-1%	-5%	-3%
	percent change in post-caustic soda consumption	-1%	-1%	-3%	-1%
	percent change in treatment costs	-2%	-1%	-3%	-3%
6.	average filter applied turbidity	0.64	0.17	1.22	1.38
	average finished turbidity	0.06	0.07	0.03	0.05
25% decrease in solids	average filter run time (hours)	122	145	54	34
	percent change in backwash volume pumped	-4%	-6%	-1%	-18%
15% decrease in organics	percent change in average finished TTHM	-7%	-10%	8%	-12%
	percent change in average finished HAA5	-10%	-10%	-13%	-17%
	percent change in pre-oxidant consumption *	-2%	-11%	-2%	-2%
	percent change in coagulant consumption**	-5%	-2%	-8%	-11%
	percent change in sludge production	-15%	-6%	-11%	-18%
	percent change in post-Cl2 consumption	0%	-4%	0%	0%
	percent change in pre-caustic soda consumption	-6%	-3%	-14%	-8%
	percent change in post-caustic soda consumption	-2%	-1%	-13%	-2%
	percent change in treatment costs	-6%	-5%	-8%	-8%
7.	average filter applied turbidity	0.61	0.17	1.21	1.24
	average finished turbidity	0.06	0.07	0.03	0.04
40% decrease in solids	average filter run time (hours)	128	150	55	38
	percent change in backwash volume pumped	-7%	-10%	-2%	-26%
20% decrease in organics	percent change in average finished TTHM	-9%	-13%	6%	-16%
	percent change in average finished HAA5	-12%	-14%	-17%	-22%
	percent change in pre-oxidant consumption *	-2%	-11%	-2%	-2%
	percent change in coagulant consumption**	-8%	-3%	-12%	-15%
	percent change in sludge production	-23%	-9%	-18%	-27%
	percent change in post-Cl2 consumption	0%	-5%	0%	0%
	percent change in pre-caustic soda consumption	-9%	-3%	-20%	-11%
	percent change in post-caustic soda consumption	-3%	-2%	-18%	-2%
	percent change in treatment costs	-9%	-6%	-12%	-11%

(continued)

Table B.1
Summary of WTP modeling results (continued)

Scenario	Parameter	Corbalis	Millwood	Muncie	Otisco
8.	average filter applied turbidity	0.58	0.15	1.27	1.04
	average finished turbidity	0.06	0.07	0.03	0.04
50% decrease in solids	average filter run time (hours)	136	161	51	45
	percent change in backwash volume pumped	-13%	-17%	3%	-38%
33% decrease in organics	percent change in average finished TTHM	-16%	-23%	-2%	-27%
	percent change in average finished HAA5	-22%	-24%	-29%	-36%
	percent change in pre-oxidant consumption *	-4%	-11%	-4%	-4%
	percent change in coagulant consumption**	-11%	-4%	-20%	-24%
	percent change in sludge production	-30%	-16%	-25%	-39%
	percent change in post-Cl2 consumption	0%	-8%	0%	0%
	percent change in pre-caustic soda consumption	-13%	-5%	-31%	-18%
	percent change in post-caustic soda consumption	-4%	-3%	-28%	-3%
	percent change in treatment costs	-13%	-7%	-18%	-17%

* Pre-oxidant is ozone for Millwood and chlorine for Corbalis, Muncie, and Otisco

** Coagulant is ferric chloride for Muncie and PACl for Corbalis, Millwood, and Otisco

REFERENCES

Adin, A., and M. Rebhun. 1977. A model to predict concentration and head-loss profiles in filtration. *Journal AWWA*, 69(8):444-453.

Ainsworth, R. 1979. *The $CaCO_3$-CO_2-H_2O system: a review of the thermodynamic data and the calculation of parameters which affect corrosion rates and scale deposition in water supplies*. WRc Report LR1009. Swindon, UK: Water Research Centre (WRc).

Ainsworth, R., and C. Jones. 1979. *A method for predicting the effects of treatment processes that are designed to increase pH, alkalinity and hardness.* WRc Report LR1048. Swindon, UK: Water Research Centre (WRc).

Alexander, R.B. 12/21/2001. Personal communication.

Alexander, R.B., R.A. Smith, M.J. Focazio, and M.A. Horn. 1999. *Source-Area Characteristics of Large Public Surface-Water Supplies in the Conterminous United States.* USGS Open-File Report 99-248. Washington DC: USGS.

Amirtharajah, A. 1985. The interface between filtration and backwashing. *Water Research*, 19(5):581-588.

Amy, G. L., Thompson, J.M., Tan, L., Davis, M.K., and Krasner, S.W. 1990. Evaluation of THM precursor contributions from agricultural drains. *Journal AWWA*, 82:1:57.

Amy, G.L., P.A. Chadik, and Z.K. Chowdhury. 1987. Developing models for predicting trihalomethane formation potential and kinetics. *Journal AWWA*, 79(7):89-97.

Arnold, J.G., R. Srinivasan, R.S. Muttiah, and J.R. Williams. 1998. Large area hydrologic modeling and assessment Part I: Model development. *Journal of the American Water Resources Association*, 34(1):73-89.

Arora, K. S.K. Mickelson, J.L. Baker, D.P. Tierney, and C.J. Peter CJ. 1996. Herbicide retention by vegetative buffer strips from runoff under natural rainfall. *Transactions of the American Society of Agricultural Engineers*, 30(6):2155-2162.

ASAE (American Society of Agricultural Engineers). 1999. *Manure Production and Characteristics.* ASAE Standard D384.1 DEC99. St. Joseph, Mich.: ASAE.

AWWA (American Water Works Association). 1996. *WATER:\STATS Water Utility Database.* Denver, Colo.: AWWA.

AWWA (American Water Works Association). 1999. *Water Quality and Treatment.* 5th ed. Letterman, R.D., ed. New York: McGraw-Hill.

Barbe, D.E., J.F. Cruise, and X. Mo. 1996. Modeling the buildup and washoff of pollutants on urban watersheds. *Water Resources Bulletin*, 32(3):511-519.

Binder, D. 2002. *Big Walnut Creek watershed runoff reduction project.* Presented at Ag-Water Alliance Expert Workshop, Indianapolis, Ind., July 17-18, 2002.

Beaulac, M.N. and K.H. Reckhow. 1982. An examination of land use and nutrient export relationships. Water Resources Bulletin, 18:1013-1024.

Borzoi, F., and T.A. Dillaha. 1996. ANSWERS-2000: Runoff and sediment transport model. *Journal of Environmental Engineering*, 122(6):493-502.

Brown, D.G., L. Bian, and S.J. Walsh. 1993. Response of a distributed watershed erosion model to variations in input data aggregation levels. *Computers and Geosciences*, 19(4):499-509.

Burkholder, J.M., M.A. Mallin, H.B. Glasgow, Jr., L.M. Larsen, M.R. McIver, G. C. Shank, N. Deamer-Melia, D.S. Briley, J. Springer, B.W. Touchette, and E.K. Hannon. 1997. Impacts to a coastal river and estuary from rupture of a large swine waste holding lagoon. *Journal of Environmental Quality*, 26(6):1451-1466.

Capel, P.D., T.A. Winterstein, and S.J. Larson. 2001. *Selected Data From Field Studies of Pesticide Runoff to Surface Waters*. USGS Water-Resources Investigations Report 00-4204. Sacramento, CA. USGS.

Caraco, D.S. 2001a. Unpublished Report on Modifications to the Watershed Treatment Model. 5/3/2001. Ellicott City, Md.: Center for Watershed Protection.

Caraco, D. 2001b. *The Watershed Treatment Model: Version 3.0*. Ellicott City, Md.: Center for Watershed Protection.

Caraco, D.S. 2000. Dynamics of Urban Stream Channel Enlargement. In *The Practice of Watershed Protection*. Edited by T.R. Schueler and H.K. Holland. Ellicott City, Md.: Center for Watershed Protection.

CDC (Centers for Disease Control and Prevention). 1995. Assessing the public health threat associated with waterborne cryptosporidiosis: report of a workshop. *Morbidity and Mortality Weekly Report* 44(No. RR-6). Atlanta, Ga: CDC.

Clausen, J.C., W.E. Jokela, F.I. Potter III, and J.W. Williams. 1996. Paired watershed comparison of tillage effects on runoff, sediment, and pesticide losses. *Journal of Environmental Quality*, 25:1000-1007.

Committee on Long-Range Soil and Water Conservation Policy, National Research Council. 1993. *Soil and Water Quality: An Agenda for Agriculture*. Washington, DC: National Academy Press.

Connolly, J.P., and R.B. Coffin. 1995. Model of carbon cycling in planktonic food webs. *Journal of Environmental Engineering-ASCE*, 121(10):682-690.

Cooke, G.D., and R.E. Carlson. 1989. *Reservoir Management for Water Quality and THM Precursor Control*. Denver, Colo.: AwwaRF and AWWA.

Cooper, J.R., J.W. Gilliam, R.B. Daniels, and W.P. Robarge. 1987. Riparian areas as filters for agricultural sediment. *Soil Science Society of America Journal*, 51:416-420.

Cooper, R.L., T.E. Stoker, L. Tyrey, J.M. Goldman, and W.K. McElroy. 2000. Atrazine disrupts the hypothalamic control of pituitary-ovarian function. *Toxicological Sciences*, 53(2):297-307.

Craun, G., Hubbs, S.A., Frost, F., Calderon, R.L., and Via, S.H. 1998. Waterborne outbreaks of cryptosporidiosis. *Journal AWWA*, 90:9:81.

Crawford, N.H., and R.K. Lindsey. 1976. *Digital Simulation Hydrology: The Stanford Watershed Model IV*. Technical Report No. 39. Palo Alto, Calif.: Stanford University.

CTIC (Conservation Technology Information Center). 1999. *CORE4 Conservation Practices Training Guide*. W. Lafayette, Ind.: CTIC.

Daniels R.B., and J.W. Gilliam. 1996. Sediment and chemical load reduction by grass and riparian filters. *Soil Science Society of America Journal*, 60:246-251.

Dharmappa, H.B., S. Vigneswaran, J. Verink, and O. Fujiwara. 1994. Water-treatment system-design for turbidity removal. I. Simulation. *Journal Of Environmental Engineering-ASCE*, 120(4):900-920.

Dillaha T.A., and S. P. Inamdar. 1997. Buffer zones as sediment traps or sources. In: *Buffer Zones: Their Processes and Potential in Water Protection*. N. Haycock, T. Burt, K. Goulding and G. Pinay (eds). Harpenden, U.K.: Quest Environmental.

Dillaha T.A., J. Pease, D. Bosch, G. Yagow, and D. Kibler. 2000. *The BMP Effectiveness Database*. Prepared for the Nutrient Task Force Sub-Committee of the USEPA Chesapeake Bay Program. Blacksburg, Va.: Virginia Polytechnic Institute and State University.

Dillaha T.A., J.H. Sherrard, D. Lee, S. Mostaghimi, and V.O. Shanholtz. 1988. Evaluation of vegetative filter strips as a best management practice for feed lots. *Journal of the Water Pollution Control Federation*, 60(7):1231-1238.

Dillaha, T.A. 1990. Role of best management practices in restoring the health of the Chesapeake Bay. In: *Perspectives on the Chesapeake Bay, 1990: Advances in Estuarine Sciences*. EPA-CBP-TRS41-90. Washington, DC: USEPA Chesapeake Bay Program.

Dillaha, T.A., J.H. Sherrard, and D. Lee. 1989. Long-term effectiveness of vegetative filter strips. *Water Environment and Technology*, 1: 418-421.

DMWW (Des Moines Water Works). 2002. Online at: www.dmww.com.

Dunne, T., and L.B. Leopold. 1978. *Water in Environmental Planning*. San Francisco: W.H. Freeman Co.

Edwards, M. 1997. Predicting DOC removal during enhanced coagulation. *Journal AWWA*, 89(5):78-89.

Edzwald, J.K., Becker, W.C., and Wattier, K. 1985. Surrogate parameters for monitoring organic matter and THM precursors. *Journal AWWA*, 77:4:172.

Enviromega. 2000. *WatPro: Inactivation/Disinfectant By-Product Model for Water Treatment Facilities, Version 2.16*. Flamborough, Ont.: Enviromega, Inc.

Fawcett, R.S., B.R. Christensen, and D.P. Tierney. 1994. The impact of conservation tillage on pesticide runoff into surface water: a review and analysis. *Journal of Soil and Water Conservation*, 49:126-135.

FHA (Federal Highway Administration). 1998. *Highway Statistics 1998*. Washington, DC: FHA, Office of Highway Policy Information.

FISRWG (Federal Interagency Stream Restoration Working Group). 1998. *Stream Corridor Restoration: Principles, Processes, and Practices*. Washington, DC: FISRWG.

FitzHugh, T.W., and D.S. Mackay. 2000. Impacts of input parameter spatial aggregation on an agricultural nonpoint source pollution model. *Journal of Hydrology*, 236:35-53.

Friedlander, S.K. 1977. *Smoke, Dust, and Haze: Fundamentals of Aerosol Behavior*. New York: John Wiley and Sons.

Garber, L.P., Salman, M.D., Hurd, H.S., Keefe, T., and Schlater, J.L. 1994. Potential risk factors for *Cryptosporidium* infection in dairy calves. *Journal of American Veterinary Medicine Association*, 205:86-91.

Gilliom, R.J., J.E. Barbash, D.W. Kolpin, and S.J. Larson. 1999. Testing water quality for pesticide pollution. *Environmental Science and Technology*, 33(7):164-169.

Giorgini, A., and F. Zingales. 1986. *Agricultural Nonpoint Source Pollution: Model Selection and Application*. New York: Elsevier.

Glasgow, H.B., and J.M. Burkholder. 2000. Water quality trends and management implications from a five-year study of a eutrophic estuary. *Ecological Applications*, 10(4):1024-1046.

Goolsby, D.A., W.A. Battaglin, G.B. Lawrence, R.S. Artz, B.T. Aulenbach, R.P. Hooper, D.R. Keeney, and G.J. Stensland. 1999. *Flux and sources of nutrients in the Mississippi-Atchafalaya River Basin: Topic 3 Report for the Integrated Assessment on Hypoxia in the Gulf of Mexico*. NOAA Coastal Ocean Program Decision Analysis Series No. 17. Silver Spring, MD: NOAA Coastal Ocean Program.

Graczyk, T.K., Evans, B.M., Shiff, C.J., Karreman, H.J., and Patz, J.A. 2000. Environmental and geographical factors contributing to watershed contamination with Cryptosporidium parvum oocysts. *Environmental Research*, 82(3):263-271.

Haith, D.A., and L.L. Shoemaker. 1987. Generalized watershed loading functions for stream flow nutrients. *Water Resources Bulletin*, 23(3):471-478.

Hall, J.K., N.L. Hartwig, and L.D. Hoffman. 1983. Application mode and alternate cropping effects on atrazine losses from a hillside. *Journal of Environmental Quality*, 12(3):336-340.

Hao, Y.L., R. Lal, R.C. Izaurralde, J.C. Ritchie, L.B. Owens, and D.L. Hothem. 2001. Historic assessment of agricultural impacts on soil and soil organic carbon erosion in an Ohio watershed. *Soil Science*, 166(2):116-126.

Harrington, G.W., Z.K. Chowdhury, and D.M. Owen. 1992. Developing a computer model to simulate DBP formation during water treatment. *Journal AWWA*, 84(11):78-86.

Hart, J., and P.R. Carlile. 1992. *Pesticide removal from lowland surface water - evaluation of different water treatment processes.* WRc Report UM1305. Swindon, UK: Water Research Centre (WRc).

Havens, K.E., E.G. Flaig, R.T. James, S. Lostal, and D. Muszick. 1996. Results of a program to control phosphorus discharges from dairy operations in South-Central Florida, USA. *Environmental Management*, 20(4):585-593.

HEC (Hydrologic Engineering Center). 1977. *Storage, Treatment, Overflow, Runoff Model, STORM, User's Manual.* Davis, Calif.: U.S. Army Corps of Engineers, Hydrologic Engineering Center.

Heiskary, S.A., and W.W. Walker. 1995. Establishing a chlorophyll-a goal for a run-of-the-river reservoir. *Lake and Reservoir Management*, 11(1):67-76.

Hill, V.R., and M.D. Sobsey. 1998. Microbial indicator reductions in alternative treatment systems for swine wastewater. *Water Science and Technology*, 38(12):119-122.

Holechek, J.L., R.D. Pieper, and C.H. Herbel. 2001. *Range Management: Principles and Practices, 4th ed.* Upper Saddle River, N.J.: Prentice-Hall.

Hom, L.W. 1972. Kinetics of chlorine disinfection in an ecosystem. *Journal of the Sanitary Engineering Division-ASCE*, 98:183-194.

Iowa DNR. 1998. *Iowa Unified Watershed Assessment, Restoration Priorities, and Restoration Action Strategy.* Des Moines, Iowa: Iowa Department of Natural Resources.

Jackson, L.L., D.R. Keeney, and E.M. Gilbert. 2000. Swine manure management plans in north-central Iowa: Nutrient loading and policy implications. *Journal of Soil and Water Conservation*, 55(2):205-212.

Jenkins, M.B., M.J. Walker, D.D. Bowman, L.C. Anthony, and W.C. Ghiorse. 1999. Use of a sentinel system for field measurements of Cryptosporidium parvum oocyst inactivation in soil and animal waste. Applied and Environmental Microbiology 65(5): 1998-2005.

Johnes, P.J. 1996. Evaluation and management of the impact of land use change on the nitrogen and phosphorus load delivered to surface waters: the export coefficient modelling approach. *Journal of Hydrology*, 183:323-349.

Jolankai, G., J. Panuska, and W. Rast. 1999. Modelling of nonpoint source pollutant loads. In *Assessment and Control of Nonpoint Source Pollution of Aquatic Ecosystems: A Practical Approach.* Edited by J.A. Thornton et al. Vol. 23, Man and the Biosphere Series. Paris: UNESCO.

Jordan, T.E., D.L. Correll, and D.E. Weller. 1993. Nutrient interception by a riparian forest receiving inputs from adjacent cropland. *Journal of Environmental Quality*, 22(3):467-473.

Kadlec, R.H., and R.L. Knight. 1996. *Treatment Wetlands*. Boca Raton, Fla.: Lewis Publishers.

Kellogg, R.L., C.H. Lander, D.C. Moffitt, and N. Gollehon. 2000. *Manure Nutrients Relative to the Capacity of Cropland and Pastureland to Assimilate Nutrients*. Publication NPS 00-0579. Washington, DC: U.S. Department of Agriculture, Natural Resources Conservation Service, Economic Research Service.

Koelmans, A.A., A. Van der Heijde, L. M. Knijff, and R. H. Aalderink. 2001. Integrated modelling of eutrophication and organic contaminant fate and effects in aquatic ecosystems. a review. *Water Research*, 35(15):3517-3536.

Lagvankar, A.L., and R.S. Gemmell. 1968. A size-density relationship for flocs. *Journal AWWA*, 60(9):1040-1046.

Larson, S.J, R.J. Gilliom, and P.D. Capel. 1999. *Pesticides in Streams: Initial Results from the National Water Quality Assessment Program*. USGS Water-Resources Investigations Report 98-4222. Washington, DC: USGS.

Larson, S.J. 2001. *Pesticides in Streams: Summary Statistics; Results of the National Water Quality Assessment Program (NAWQA), 1992-1998*. Washington DC: USGS.

Lawler, D.F., C.R. O'Melia, and J.E. Tobiason. 1980. Integral water treatment plant design: from particle size to plant performance. In *Particulates in Water*. Edited by C. Kavanaugh and J.O. Leckie. Washington, DC: American Chemical Society.

LeChevallier, M.W., and W.D. Norton. 1995. *Giardia* and *Cryptosporidium* in raw and finished water. *Journal AWWA*, 87(9):54-68.

Lee, K.H., T.M. Isenhart, R.C. Schultz, and S.K. Mickelson. 1998. Nutrient and sediment removal by switchgrass and cool-season grass filter strips in Central Iowa, USA. *Agroforestry Systems*, 44(2- 3):121-132.

Line, D.E., D.L. Osmond, R.A. McLaughlin, W.A. Harman, K.L. Tweedy, L.A. Lombardo, and J. Spooner. 2000a. Nonpoint sources. *Water Environment Research*, 72(5):1054-1069.

Line, D.E., W.A. Harman, G.D. Jennings, E.J. Thompson, and D.L. Osmond. 2000b. Nonpoint-source pollutant load reductions associated with livestock exclusion. *Journal of Environmental Quality*, 29(6):1882-1890.

Litke, D.W. 1999. *Review of Phosphorus Control Measures in the United States and their Effects on Water Quality*. USGS Water-Resources Investigations Report 99-4007. Washington, DC: USGS.

Lowrance R., L.S. Altier, J.D. Newbold, R.R. Schnabel, P.M. Groffman, J.M. Denver, D.L. Correll, J.W. Gilliam, J.L. Robinson, R.B. Brinsfield, W. Lucas, and A.H. Todd. 1997. Water quality functions of riparian forest buffers in Chesapeake Bay watersheds. *Environmental Management*, 21(5):687-712.

Lowrance, R., R.L. Todd, J. Fail Jr., O. Hendrickson Jr., R. Leonard, and L. Asmussen. 1984. Riparian Forests As Nutrient Filters In Agricultural Watersheds. BioScience 34(6):374-377.

LSWRPC (Lake Springfield Watershed Resources Planning Committee). 1998. *Enhancing Water Quality*. Springfield, Ill.: Sangamon County Soil and Water Conservation District.

Lynch, L., and R. Tjaden. 2000. *When a Landowner Adopts a Riparian Buffer–Benefits and Costs*. University of Maryland Cooperative Extension, Fact Sheet 774. College Park, Md.: University of Maryland.

Mallevialle, J., and I.H. Suffet. 1987. *Identification and Treatment of Tastes and Odors in Drinking Water*. Denver, Colo.: AwwaRF and AWWA.

Malley, J.P., and J.K. Edzwald. 1991. Concepts for dissolved-air flotation treatment of drinking water. *Journal of Water Supply Research and Technology - Aqua*, 40(1):7-17.

Mallin, M.A., J.M. Burkholder, M.R. McIver, G.S. Shank, H. B. Glasgow, Jr., B.W. Touchette, and J. Springer. 1997. Comparative effects of poultry and swine waste lagoon spills on the quality of receiving streamwaters. *Journal of Environmental Quality*, 26(6):1622-1631.

Marquez, C.O., C.A. Cambardella, T.M. Isenhart, and R.C. Schultz. 1999. Assessing soil quality in a riparian buffer strip system by testing organic matter fractions. Agroforestry Systems, 44: 133-140.

Mattikalli, N.M., and K.S. Richards. 1996. Estimation of surface water quality changes in response to land use change: application of the export coefficient model using remote sensing and geographical information system. *Journal of Environmental Management*, 48:263-282.

McDowell, R.W., A.N. Sharpley, L.M. Condron, P.M. Haygarth, and P.C. Brookes. 2001. Processes controlling soil phosphorus release to runoff and implications for agricultural management. *Nutrient Cycling in Agroecosystems*, 59(3):269-284.

Meals, D., R. Hopkins, S. Fiske, and R. Langdon. 2001. *Lake Champlain Basin Agricultural Watersheds Section 319 National Monitoring Program Project: Final Report*. Waterbury, Vt.: Vermont Department of Environmental Conservation.

Missouri UWASC (Unified Watershed Assessment Steering Committee). 1998. *State of Missouri Unified Watershed Assessment: Final Report*. Columbia, Mo.: USDA-NRCS.

Nakao, M., B. Sohngen, L. Brown, and R. Leeds. 1999. *The Economics of Vegetative Filter Strips*. Ohio State University Extension AE-0006-99. Columbus, Ohio: Ohio State University Extension.

Natural Resources Conservation Service (NRCS). 2002. *National Handbook of Conservation Practices*. Washington, DC: NRCS.

NCDEHNR (North Carolina Department of Environment, Health and Natural Resources). 1993. *Neuse River Basinwide Water Quality Management Plan*. Raleigh, N.C.; NCDEHNR, Division of Environmental Management, Water Quality Section.

Nebraska Cooperative Extension. 1997a. *Agricultural management practices to reduce atrazine in surface water*. Institute of Agriculture and Natural Resources, Cooperative Extension Publication G96-1299-A. Lincoln, Neb.: University of Nebraska.

Nebraska Cooperative Extension. 1997b. *BMPs to reduce atrazine runoff from dryland corn and sorghum in south central and southeast Nebraska*. Institute of Agriculture and Natural Resources, Cooperative Extension Publication NF97-322. Lincoln, Neb.: University of Nebraska.

Novotny, V. and G. Chesters. 1981. Handbook of Nonpoint Source Pollution: Sources and Management. New York: Van Nostrand Reinhold Co.

Novotny, V. and H. Olem. 1994. *Water Quality: Prevention, Identification, and Management of Diffuse Pollution*. New York: Van Nostrand Reinhold.

Nydam, D.V. S.E. Wade, S.L. Schaaf, and H.O. Mohammed. 2001. Number of Cryptosporidium parvum oocysts or Giardia spp cysts shed by dairy calves after natural infection. *American Journal of Veterinary Research*, 62 (10): 1612-1615.

Olson, M.E., J. Goh, M. Phillips, N. Guselle, and T.A. McAllister. 1999. *Giardia* cyst and *Cryptosporidium* oocyst survival in water, soil, and cattle feces. *Journal of Environmental Quality*, 28(6):1991-1996.

Ongerth, J., and H. Stibbs. 1989. Prevalence of *Cryptosporidium* infection in dairy calves in western Washingon. *American Journal of Veterinary Research*, 50(7):1060 – 1070.

Oppenheimer, J.A., E.M. Aieta, R.R. Trussell, J.G. Jacangelo, and I.N. Najm. 2000. *Evaluation of Cryptosporidium Inactivation in Natural Waters.* Denver, Colo.: AwwaRF and AWWA.

Osborne, L.L., and D.A. Kovacic. 1993. Riparian vegetated buffer strips in water-quality restoration and stream management. *Freshwater Biology*, 29(2):243-258.

Owens, L.B., W.M. Edwards, and R.W. Van Keuren. 1996. Sediment losses from a pastured watershed before and after stream fencing. *Journal of Soil and Water Conservation*, 51(1):90-94.

Palone, R.S., and A.H. Todd, eds. 1998. *Chesapeake Bay riparian handbook: a guide for establishing and maintaining riparian forest buffers.* NA-TP-02-97. Radnor, Pa.: U.S. Department of Agriculture, Forest Service.

Parker, D.S., W.J. Kaufmann, and D. Jenkins. 1972. Floc breakup in turbulent flocculation processes. *Journal of the Sanitary Engineering Division-ASCE*, 98:79-99.

Peterjohn, W.T., and D.L. Correll. 1984. Nutrient dynamics in an agricultural watershed: Observations on the role of a riparian forest. *Ecology*, 65(5):1466-1475.

Pitt R. 1993. Source loading and management model (SLAMM). Presented at the National Conference on Urban Runoff Management, March 30-April 2, 1993, Chicago, IL.

Pitt, R., and J. Voorhees. 1995. Source Loading and Management Model (SLAMM). In *National Conference on Urban Runoff Management: Enhancing Urban Watershed Management at the Local, County, and State Levels (Seminar Publication).* EPA-625-R-95-003. Cincinnati, Ohio: USEPA, Center for Environmental Research Information.

PSU (Pennsylvania State University). 1992. *Agricultural BMP Effectiveness Database.* University Park, Pa.: Pennsylvania State University, Department of Agricultural and Biological Engineering.

Raschke, R.L. 1994. Phytoplankton bloom frequencies in a population of small southeastern impoundments. *Lake and Reservoir Management*, 8(2):205-210.

Reckhow, K.H., and S.C. Chapra. 1999. Modeling excessive nutrient loading in the environment. *Environmental Pollution*, 100:197-207

Renard, K.G., G.R. Foster, G.A. Weesies, and J.P. Porter. 1991. RUSLE: Revised Universal Soil Loss Equation. *Journal of Soil and Water Conservation*, 46(1):30-33.

Robertson, L.J., P.G. Smith, A.T. Grimason, and H.V. Smith. 1999. Removal and destruction of intestinal parasitic protozoans by sewage treatment processes. *International Journal of Environmental Health Research,* 9:85 – 96.

Rose, J.B. 1990. Occurrence and control of *Cryptosporidium* in drinking water. In *Drinking Water Microbiology.* Edited by G.A. McFeters. New York: Springer-Verlag.

Rose, J.B. 1997. Environmental ecology of *Cryptosporidium* and public health implications. *Annual Review of Public Health*, 18:135-161.

Rose, J.B., D.E. Huffman, K. Riley, S.R. Farrah, J.O. Lukasik, and C.L. Hamann. 2001. Reduction of enteric microorganisms at the Upper Occoquan Sewage Authority Water Reclamation Plant. *Water Environment Research*, 73(6):711-720.

Rose, J.B., L.J. Dickson, S.R. Farrah, and R.P. Carnahan. 1996. Removal of pathogenic and indicator microorganisms by a full-scale water reclamation facility. *Water Research*, 30(11):2785-2797.

Rudra, R.P., W.T. Dickinson, and E.L. von Euw. 1993. Importance of precise rainfall inputs in nonpoint source pollution modeling. *Transactions of the American Society of Agricultural Engineers*, 36(2):445-450.

Saatci, A.M., and C.S. Oulman. 1980. The bed depth service time design method for deep bed filtration. *Journal AWWA*, 72(9):524-528.

San Diego-McGlone, M.L., S.V. Smith, and V. Nicolas. 2000. Stoichiometric interpretations of C:N:P ratios in organic waste materials. *Marine Pollution Bulletin*, 40(4):325-330.

Sattar, S.A., C. Chauret, V.S. Springthorpe, D. Battigelli, M. Abbaszadegan, and M. LeChevallier. 1999. *Giardia Cyst and Cryptosporidium Oocyst Survival in Watersheds and Factors Affecting Inactivation*. Denver, Colo.: AwwaRF and AWWA.

SAWPA. 2001. Santa Ana Watershed Project Authority. Online at: www.sawpa.org.

Sawyer, C.H. 1947. Fertilization of lakes by agricultural and urban drainage. *New England Water Works Association*, 61:109-127.

Schmitt, T.J., M.G. Dosskey, and K.D. Hoagland. 1999. Filter strip performance and processes for different vegetation, widths, and contaminants. *Journal of Environmental Quality* 28(5):1479-1489.

Schueler, T.R. 1987. *Controlling Urban Runoff: A Practical Manual for Planning and Designing Urban Best Management Practices*. Document No. 87703. Washington, DC: MWCOG (Metropolitan Washington Council of Governments).

Schueler, T.R. 2000. The importance of imperviousness. In *The Practice of Watershed Protection*. Edited by T.R. Schueler and H.K. Holland. Ellicott City, Md.: Center for Watershed Protection.

Schwer C.B., and J.C. Clausen. 1989. Vegetative filter treatment of dairy milkhouse wastewater. *Journal of Environmental Quality*, 18:446-451.

Shepard, R. 2000. Nitrogen and phosphorous management on Wisconsin farms: Lessons learned for agricultural water quality programs. *Journal of Soil and Water Conservation*, 55(1):63-68.

Shepherd, D., and J. Hart. 1993. *Ozone contactor design*. WRc Report UM1402. Swindon, UK: Water Research Centre (WRc).

Sischo, W.M., E.R. Atwill, L.E. Lanyon, and J. George. 2000. *Cryptosporidia* on dairy farms and the role these farms may have in contaminating surface water supplies in the northeastern United States. *Preventive Veterinary Medicine*, 43(4):253-267.

Smith, C.M. 1992. Riparian afforestation effects on water yields and water quality in pasture catchments. *Journal of Environmental Quality*, 21(2):237-245.

Smith, R.A., G.E. Schwarz, and R.B. Alexander. 1997. Regional interpretation of water-quality monitoring data. *Water Resources Research*, 33(12): 2781-2798.

Smoluchowski, M. 1917. Vesuch einer mathematischen theorie der koagulations – kinetik kolloider losungen. *Zeitschrift fur Physikalische Chemie*, 92(2):129-168.

Sovell, L.A., B. Vondracek, J.A. Frost, and K.G. Mumford. 2000. Impacts of rotational grazing and riparian buffers on physicochemical and biological characteristics of southeastern Minnesota, USA, streams. *Environmental Management*, 26(6):629-641.

Srivastava, P., D.R. Edwards, T.C. Daniel, P.A. Moore, and T.A. Costello. 1996. Performance of vegetative filter strips with varying pollutant source and filter strip lengths. *Transactions of the American Society of Agricultural Engineers*, 39(6):2231-2239.

Srivastava, P., T.A. Costello, D.R. Edwards, and J.A. Ferguson. 1998. Validating a buffer zone nutrient transport model for variable poultry litter-length. *Transactions of the American Society of Agricultural Engineers*, 41(1):89-95

Storm, D.E., M. White, M.D. Smolen, and H. Zhang. 2001. *Modeling Phosphorous Loading for the Lake Eucha Basin: Draft Report*. Submitted to Tulsa Metropolitan Utility Authority. Stillwater, Okla.: Oklahoma State University.

Summers, R.S, Z.K. Chowdhury, G. Solarik, W. Swanson, J. Sohn, and G.L. Amy. 1999. Extensions and Verification of the Water Treatment Plant Model. In *Proceedings of the 1999 AWWA Annual Conference and Exposition*. Denver, Colo.: AWWA.

Tetra Tech. 2001. *Spreadsheet Tool for the Estimation of Pollutant Load (STEPL), Version 2.0.* Fairfax, Va.: Tetra Tech, Inc.

Thomas, D.N, S.J. Judd, and N. Fawcett. 1999. Flocculation modelling: a review. *Water Research*, 33(7):1579-1592.

Tiedemann, A.R., D.A. Higgins, T.M. Quigley, H.R. Sanderson, and D.B. Marx. 1987. Responses of fecal coliform in streamwater to four grazing strategies. *Journal of Range Management*, 40:322-329.

Tseng T., B.D. Segal, and M. Edwards. 2000. Increasing alkalinity to reduce turbidity. *Journal AWWA*, 92(6):44-54.

Tseng, T., and M. Edwards. 1999. Predicting full-scale TOC removal. *Journal AWWA*, 91(4):159-170.

U.S. Census Bureau. 2000. "Table 1. Land Area, Population, and Density for States and Counties: 1990." Washington, D.C.: United States Department of Commerce, U.S. Census Bureau, Population Division, Population & Housing Programs Branch.

USDA. 2000. *Draft Comprehensive Nutrient Management Planning Technical Guidance*. National Planning Procedures Handbook, Subpart E, Parts 600.50-600.54 and Subpart F, Part 600.75. Washington, DC: USDA, Natural Resources Conservation Service.

USEPA. 1976. *Loading Functions for Assessment of Water Pollution for Non-Point Sources*. EPA 600-2-76-151. Washington, DC: USEPA.

USEPA. 1985. *Water quality assessment: A screening procedure for toxic and conventional pollutants in surface and ground water. Part 1*. EPA-600-6-85-002a. Athens, Ga.: USEPA, Office of Research and Development, Environmental Research Laboratory.

USEPA. 1988. *Storm Water Management Model User's Manual, Version 4*. EPA-600-3-88-001a. Athens, Ga.: USEPA, Office of Research and Development, Environmental Research Laboratory.

USEPA. 1991. *Modeling of Nonpoint Source Water Quality in Urban and Non-Urban Areas*. EPA 600-3-91-039. Athens, Ga.: USEPA, Office of Research and Development, Environmental Research Laboratory.

USEPA. 1993. *Guidance Specifying Management Measures for Sources of Nonpoint Pollution in Coastal Waters*. EPA-840-B-93-001c. Washington, DC: USEPA, Office of Water.

USEPA. 1994a. *Section 319 Nonpoint Source Success Stories: Volume I*. EPA 841-S-94-004. Washington, DC: USEPA, Office of Wetlands, Oceans and Watersheds, Assessment and Watershed Protection Division, Nonpoint Source Control Branch.

USEPA. 1994b. *Water Treatment Plant Simulation Program Version 1.55.* EPA 814-C-93-001. Washington, DC: USEPA.

USEPA. 1997a. *Compendium of Tools for Watershed Assessment and TMDL Development.* EPA 841-B-97-006. Washington, DC: USEPA, Office of Wetlands, Ocean, and Watersheds, Assessment and Watershed Protection Division.

USEPA. 1997b. *Hydrological Simulation Program - FORTRAN (HSPF): User's Manual for Version 11.0.* EPA 600-R-97-080. Athens, Ga.: USEPA, Office of Research and Development, Environmental Research Laboratory.

USEPA. 1998. *Better Assessment Science Integrating Point and Nonpoint Sources*. EPA-823-R-98-006. Washington, DC: USEPA, Office of Water, Exposure Assessment Branch.

USEPA. 1999a. *Indicators of the Environmental Impacts of Transportation, 2nd ed.* EPA 230-R-99-001. Washington, DC: USEPA, Office of Policy, Planning, and Evaluation.

USEPA. 1999b. *Protocol for Developing Nutrient TMDLs*. EPA 841B-99-007. Washington, DC: USEPA, Office of Wetlands, Oceans, and Watersheds.

USEPA. 2000a. *National Management Measures to Control Nonpoint Source Pollution from Agriculture – Draft.* Washington, DC: USEPA, Office of Water, Nonpoint Source Control Branch.

USEPA. 2000b. *National Water Quality Inventory: 1998 Report.* EPA-841-F-00-006. Washington, DC: USEPA, Office of Water.

USEPA. 2001a. *Development Document for the Proposed Revisions to the National Pollutant Discharge Elimination System Regulation and the Effluent Guidelines for Concentrated Animal Feeding Operations.* EPA-821-R-01-003. Washington DC: USEPA, Office of Water.

USEPA. 2001b. *Drinking water exposure assessment for atrazine and various chloro-triazine and hydroxy-triazine degradates.* Washington DC: USEPA, Office of Pesticide Programs, Environmental Fate and Effects Division.

USEPA. 2001c. *Environmental and Economic Benefit Analysis of Proposed Revisions to the National Pollutant Discharge Elimination System Regulation and the Effluent Guidelines for Concentrated Animal Feeding Operations.* EPA-821-R-01-002. Washington DC: USEPA, Office of Water.

USEPA. 2001d. *Environmental Assessment of Proposed Revisions to the National Pollutant Discharge Elimination System Regulation and the Effluent Guidelines for Concentrated Animal Feeding Operations.* EPA-821-R-01-001. Washington DC: USEPA, Office of Water.

USEPA. 2001e. *Long Term 2 Enhanced Surface Water Treatment Rule- Draft Preamble.* Washington, DC: USEPA, Office of Ground Water and Drinking Water.

USEPA. 2001f. *Proposed revisions to the NPDES Permit Regulations and the Effluent Guidelines and Standards for Concentrated Animal Feeding Operations.* January 12, 2001, 66 Federal Register 2959.

USEPA. 2002a. 40 CFR Parts 9, 122 and 412. National Pollutant Discharge Elimination System Permit Regulation and Effluent Limitation Guidelines and Standards for Concentrated Animal Feeding Operations; Final Rule.

USEPA. 2002b. *National Water Quality Inventory: 2000 Report.* EPA-841-R-02-001. Washington, DC: USEPA, Office of Water.

USGS. 1999. *The Quality of Our Nation's Waters–Nutrients and Pesticides.* Circular 1225. Reston, Va.: USGS.

USGS. 2000. *Data on nutrients in the streams, rivers, and ground water of the United States.* Reston, Va.: USGS.

USGS. 2001. *Selected Findings and Current Perspectives on Urban and Agricultural Water Quality by the National Water-Quality Assessment Program.* Fact Sheet 047-01. Reston, Va.: USGS.

Uusi-Kamppa, J., E. Turtola, H. Hartikainen, and T. Ylaranta. 1997. The interactions of buffer zones and phosphorus runoff. In: *Buffer Zones: Their Processes and Potential in Water Protection.* N. Haycock, T. Burt, K. Goulding and G. Pinay (eds). Harpenden, U.K.: Quest Environmental.

Vieux, B.E., and S. Needham. 1993. Nonpoint-pollution model sensitivity to grid-cell size. *Journal of Water Resources Planning and Management*, 119(2):141-157.

Vollenweider, R.A. 1968. *Scientific fundamentals of the eutrophication of lakes and flowing waters, with particular reference to nitrogen and phosphorus as factors in eutrophication.* DAS/CSI/68.27. Paris: Organization of Economic Cooperation and Development.

Walker, J.F., S.A. Pickard, and W.C. Sonzogni. 1989. Spreadsheet watershed modeling for nonpoint-source pollution management in a wisconsin basin. *Water Resources Bulletin*, 25(1):139-147.

Walker, W.W. 1983. Significance of eutrophication in water supply reservoirs. *Journal AWWA*, 75(1):38-42.

Walker, W.W., and K.E. Havens. 1995. Relating algal bloom frequencies to phosphorus concentrations in Lake Okeechobee. *Lake and Reservoir Management*, 11(1):77-83.

Wauchope, R. D., T. M. Buttler, A. G. Hornsby, P. M. Augustijn-Beckers, and J. P. Burt. 1992. The SCS/ARS/CES pesticide properties database for environmental decision making. *Reviews of Environmental Contamination and Toxicology* 123:1-155.

Whittemore, R.C., and J. Beebe. 2000. EPA's BASINS model: good science or serendipitous modeling? *Journal of the American Water Resources Association*, 36(3):493-499.

Williamson, R.B., C.M. Smith, and A.B. Cooper. 1996. Watershed riparian management and its benefits to a eutrophic lake. *Journal of Water Resources Planning and Management - ASCE*, 12

Wischmeier, W. H. 1976. Use and Misuse of the Universal Soil Loss Equation. Journal of Soil and Water Conservation, 31(1): 5-9.2(1):24-33.

WRc/D&B. 2001. *Otter, Version 2.1.2.* Swindon, UK:WRc/D&B, LLC.

Young, R.A., C.A. Onstad, D.D. Bosch, and W.P. Anderson. 1989. AGNPS: a nonpoint-source pollution model for evaluating agricultural watersheds. *Journal of Soil and Water Conservation*, 44(2):168-173.

Yu, R.F, S.F. Kang, S.L. Liaw, and M.C. Chen. 2000. Application of artificial neural network to control the coagulant dosing in water treatment plant. *Water Science and Technology*, 42(3-4):403-408.

Zhang, H.P., and K. Yamada. 1996. Estimation for urban runoff quality modeling. *Water Science and Technology*, 34(3-4):49-54.

ABBREVIATIONS

AFO	Animal Feeding Operation
AGNPS	Agricultural Non-Point Source Pollution Model
ANSWERS	Areal Nonpoint Source Watershed Environmental Response Simulation
AU	Animal unit
BASINS	Better Assessment Science Integrating Point and Nonpoint Sources
BMP	Best Management Practice
CAFO	Concentrated Animal Feeding Operation
CREAMS	Chemicals, Runoff and Erosion from Agricultural Management Systems
EPIC	Erosion/Productivity Impact Calculator
FCWA	Fairfax County Water Authority, Fairfax, Virginia
FHWA	Federal Highway Administration Model
GIS	Geographic Information System
GLEAMS	Groundwater Loading Effects of Agricultural Management
GWLF	Generalized Watershed Loading Functions
HAA5	Five Haloacetic Acids
HSPF	Hydrological Simulation Program - FORTRAN
IAWCo	Indiana American Water Company
MUSLE	Modified Universal Soil Loss Equation
NAWQA	National Water-Quality Assessment Program (USGS)
NOM	Natural Organic Matter
NPDES	National Pollutant Discharge Elimination System
NRCS	Natural Resources Conservation Service
OCWA	Onondaga County Water Authority, Syracuse, New York
QUAL2E	Enhanced Stream Water Quality Model
RUSLE	Revised Universal Soil Loss Equation
SLOSS-PHOSPH	Sediment and Phosphorus Prediction
STEPL	Spreadsheet Tool for the Estimation of Pollutant Loads (STEPL)
SWAT	Soil and Water Assessment Tool
SWIM	Soil and Water Integrated Model
SWMM	Storm Water Management Model
SWRRBWQ	Simulator for Water Resources in Rural Basins-Water Quality

TOC	Total Organic Carbon
TP	Total Phosphorus
TSS	Total Suspended Solids
TTHM	Total Trihalomethanes
USEPA	U.S. Environmental Protection Agency
USGS	U.S. Geological Survey
USLE	Universal Soil Loss Equation
WMM	Watershed Management Model
WTM	Watershed Treatment Model
WTP	Water Treatment Plant
WWTP	Wastewater Treatment Plant

IWA Publishing's authorised EU representative for General Product Safety Regulations is Diane D'Arras, 15 rue Duret, 75116 Paris, France, e-mail: safety@iwap.co.uk.

Printed and bound by CPI Group (UK) Ltd, Croydon, CR0 4YY
17/09/2026
02218813-0018

Establishing Site-Specific Flushing Velocities

The mission of the Awwa Research Foundation (AwwaRF) is to advance the science of water to improve the quality of life. Funded primarily through annual subscription payments from over 1,000 utilities, consulting firms, and manufacturers in North America and abroad, AwwaRF sponsors research on all aspects of drinking water, including supply and resources, treatment, monitoring and analysis, distribution, management, and health effects.

From its headquarters in Denver, Colorado, the AwwaRF staff directs and supports the efforts of over 700 volunteers, who are the heart of the research program. These volunteers, serving on various boards and committees, use their expertise to select and monitor research studies to benefit the entire drinking water community.

Research findings are disseminated through a number of technology transfer activities, including research reports, conferences, videotape summaries, and periodicals.